D0679654

BARRON'S ART HANDBOOKS

WATER EFFECTS

BARRON'S

BARRON'S ART HANDBOOKS

WATER EFFECTS

BARRON'S

CONTENTS

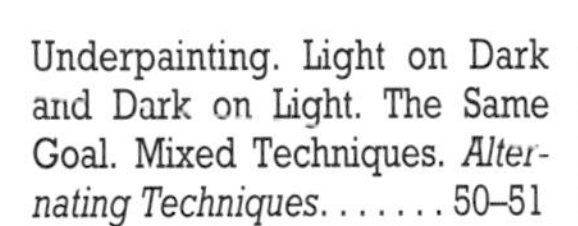

WATER AS A SUBJECT

Given the important role that water plays in nature, and its almost "constant presence in our daily life, it is not surprising that it is represented in many works of art, sometimes as the central element and sometimes as a secondary feature.

Techniques

As can be seen from these pictures, all techniques are possible for portraying water.

The Procedure for Each Technique

One can identify some or other general characteristic of each medium that makes it special.

Watercolor, for example, gives luminescent results thanks to its transparency.

Oils, on the other hand, allow one to work opaquely and with impasto or to paint using transparencies. The alternative methods of application, whether working directly *alla prima* or over several sessions, will give very different results.

Using pencils and media in stick form as tools (charcoal, conte crayon, pencils, chalks, pastels, wax crayons) allows the artist to use drawing techniques, to add color and to shade by cross-hatching.

Ink can be used as a wash applied with a brush, with a nib (which can be used for cross-hatching) or a reed pen. Wash and line techniques can be combined.

Acrylic can be used as a transparent paint or opaquely. Its fast drying time (in contrast to oil paint) means that a work can be completed in a single session.

Sketches and the Finished Work

Rapid studies in the form of sketches can be made using dry

William Turner, Thames view with barges and punt, *watercolor.*

Winslow Homer, Bermuda shoreline, *watercolor.*

Gustave Caillebotte, Vue des Toits (Effet de Neige), *oil on canvas.*

Camille Pissarro, Le Pont Boieldieu à Rouen, *oil on canvas.*

Edward Seago, Dusk light, Rouen, *watercolor.*

John Twachtman, February, *watercolor.*

Vincent van Gogh, Boats at Sainte Maries, *oil on canvas.*

Joaquín Sorolla, Pines in Valsaín *(detail).*

media (charcoal, conte crayon, chalks, pastels), wax crayons, monochromatic washes or washes using few colors, pencil, etc. Such sketches can establish the framing, composition and tonal values of the theme. Sketches are used for early problem solving regarding composition, lights and darks, size, shape, etc. Studies and sketches are made as a way of establishing elements that are going to be used and as the first steps in the execution of the finished work (particularly when the work is of larger dimensions and requires more complex solutions).

Nevertheless, there are some rapid sketches that can be considered finished masterpieces thanks to their clear definition of the theme, and because of their freshness or virtuosity.

Color and texture

When the artist approaches the subject of water, its movement and appearance must be carefully observed. It can then be represented using one medium or by mixing several. The means of representation are none other than color (and its distribution) and texture. Here one can see how a mixed technique using a combination of felt-tipped pens and ink washes can rapidly produce a very effective sketch.

Miquel Ferrón, Boats in the harbour, *mixed technique with felt-tipped pens and ink washes.*

MORE ON THIS SUBJECT

- The Initial Sketch and Tonal Values **p. 20**
- Tonal Values and Color **p. 22**
- Mixed Techniques, I **p. 24**
- Oils **p. 26**

THE PHYSICAL PROPERTIES OF WATER

Water can be found in three states: as a liquid, a solid, or a gas. It forms a part of all living organisms and is the major component of the earth's surface. Given its importance, it is an element that artists paint very frequently and need to learn to represent in any of its states.

Three States: Liquid, Solid, and Gas

Water is found covering so much of the earth's surface that it naturally drew the interest of mankind to master its navigation. As an expression of heroic actions and their mythical dimension, the representation of naval battles on rough waters is very familiar to us. So too are bucolic paintings on the banks of slow rivers or rural scenes with wild torrents. Snowy mountains or icy ponds beneath limpid or cloud-filled skies are other variations of the impressive and never-ending possibilities of nature.

It is important for a painter to understand the way water behaves in order to well represent it.

Water can be presented in any of three states. In its liquid state, it appears in seas, rivers, marshes, ponds, pools, rain, the contents of a bottle or a glass, tears, dew, and so on.

In its solid state, water appears in the form of snow or ice. Powdery snow and ice do not have the same characteristics. Stalagmites and stalactites are very different in form from frost. The shape of an iceberg, as an example of a large volume, can adopt very different forms, although they are always very identifiable.

Water vapor, either in the form of a cloud, as mist, or as steam can be difficult for a painter to represent.

Transparency, Reflection, and Refraction

When a beam of light hits a surface that has the capacity to reflect or refract it, such as is the case with water (depending on the angle of observation), it can cause reflections, multiple illuminations, reflected images, and direct images to appear.

In any of the three states, water has a capacity for transparency that must be analyzed in

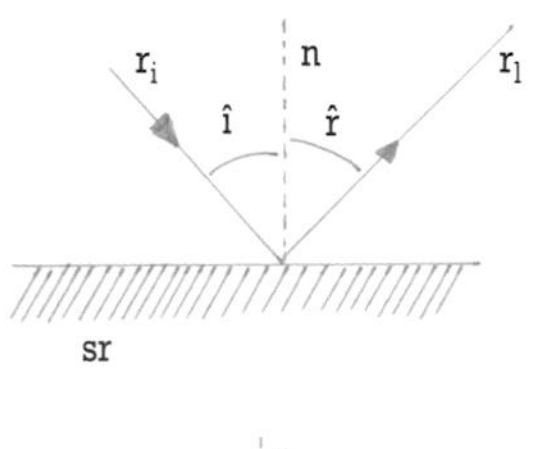

Graphic illustration of the reflection of a beam of light. The trajectory of a beam reflected by the surface on which it falls:
sr: reflective surface
n: normal or perpendicular
r_i: incidental beam
r_l: reflected beam
$\hat{i}$: angle of light, or incidence
$\hat{r}$: angle of reflection

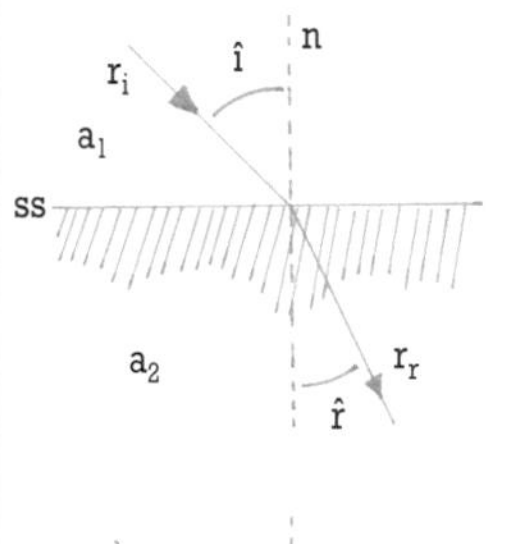

Graphic illustration of the reflection of a beam of light. The trajectory of a beam passing through the surface.
a_1: medium 1
a_2: medium 2
ss: surface of separation between media 1 and 2.
n: normal or perpendicular
r_i: incidental beam
r_r: refracted beam
$\hat{i}$: angle of light, or incidence
$\hat{r}$: angle of refraction

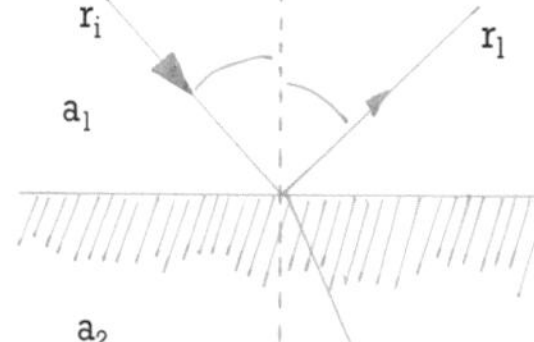

Graphic representation of a medium that causes both reflection and refraction of a beam of light.

Egon Schiele, Sailing boats on water moved by waves, *oil and pencil on card.*

The typical luminosity of a rainy day is reflected in these puddles. Vicenç Ballestar.

William Turner,
Venice: Dusk.

Alfred Sisley,
Canal Saint Martin.

MORE ON THIS SUBJECT
- Reflections **p. 16**
- Water and Light **p. 92**

Gustave Caillebotte,
Yachts at Argenteuil.

Edward Hopper, The White River, Foreground. *The atmosphere of light at dusk is reflected in the surface of the river. The reflected images of the trees on the riverbank are also visible.*

each case. For example, the water of a river containing particles in suspension usually has a specific color. However, if the water is not very turbulent, it is often possible to see the stones or mud of the riverbed. As for the sea, the sand beyond the shoreline can be seen through the water, as can rocks and seaweed that do not lie too deep.

In the case of water vapor, objects appear blurred when seen through it.

Ice is very transparent. In some cases it causes distortions of an image, sometimes even causing images to appear inverted. Ice has the capacity to cause reflection as well as complex refraction. It is water's capacity of refraction as a liquid, solid, or gas that causes its transparency.

The painter must first analyze the subject and determine its composition, i.e., where the edges or pictorial limits of a lake or river should lie. Undoubtedly, however, the most complex aspect of this study is deciding on the chromatic and textural elements most suitable for representing water, depending on the technique being used. To make these decisions, it is necessary to take into account and observe the possible reflections (of light and image) and transparencies.

Reflection and Refraction: Physical Rules

A surface such as water causes both phenomena (reflection and refraction) at the same time. Part of the incidental, that is, the falling or striking, beam of light (which divides as it enters into contact with the surface) is reflected and part is refracted. The direction of the incidental beam of light, the normal and the reflected beam are all on the same plane, as are the direction of the incidental beam, the normal and the refracted beam.

Whenever the source of light, the normal at the point of incidence, and the point of view of the painter are on the same plane, the viewer will see reflections and multiple refraction in a wide area, as occurs with a mirror, calm water or even the infinite small points of light on turbulent water.

Great Masters: Reflection, Refraction, and Transparency

William Turner, in a supremely creative watercolor, *Venice: Dusk* displays his mastery of all aspects of this medium in showing the luminescence of Venice. Spots of light, reflected images, the surface of the water and the atmosphere create an admirably resolved whole.

The atmosphere of light at dusk achieved by Edward Hopper in *The White River, Foreground* is reflected in the surface of the river. Similarly the images of the trees on the bank appear reflected as well.

What Color Is It?

The color of water in any of its states depends on its surroundings and on the light that illuminates it. When treating a given subject, the artist looks for and aims to create tonal and color areas using dominant colors in order to interrelate the different planes.

An artist's interpretation. Ceferino Oliver.

THE HORIZON AND PLANES

In order to visually understand a subject, its various elements must be situated correctly in relation to the horizon line, and drawn using correct perspective. Meanwhile from the point of view of chromatic analysis, it is useful to establish the positions of the different planes, which normally consist of the foreground, the mid-ground, and finally the background.

The Important Line. The Point of View

For any particular framing, it is essential to situate the line of the horizon on the support before beginning to paint a landscape. In practice, this is the horizontal line, either real or imaginary, found at the same height as the painter's eyes and which also describes the location of the horizon. All the space located above the line of the horizon forms the background of the painting.

The point of view in a painting shows the direction in which the artist is looking. It should be located so that it is not excessively central. It should appear on the horizon line, although displaced to the right or the left.

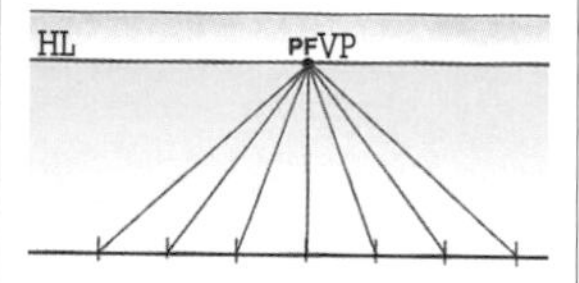

Representation of the horizon line (HL) with a vanishing point (VP), for parallel perspective. The point of view coincides with the vanishing point.

In a seascape the horizon line coincides with the horizon of the sea.

High Horizons and Low Horizons

Depending on the location from which a painter chooses to depict his or her vision of a landscape, one can talk about a high horizon or a low horizon. This question is particularly relevant when depicting seascapes. As can be seen in the illustration with a high horizon, the physical space occupied by the sea on the canvas is very large, and the perspective is more aerial.

In contrast, if the painter stands on the same level as the surface of the water, the painting will have a low horizon producing a composition with a very narrow band of sea.

The line of the horizon can even fall outside the painting. This would be an extreme case of a high horizon, with a completely aerial view. When this formula is used, the impression it causes is one of immediacy.

Perspectives: Parallel, Oblique, and Aerial

In order to represent a subject that has volume and dimension on the flat surface of your support (whether it is canvas, paper, wood or cardboard) the rules of perspective must be employed. Parallel perspective, or perspective using one vanishing point (the point on the horizon where the lines converge) maintains parallel lines, either vertically or horizontally. This is the simplest perspective.

With oblique perspective, which has two vanishing points,

The horizon line is almost at midpoint of this composition. The distance is emphasized by the figures who sit on the rock viewing the horizon.

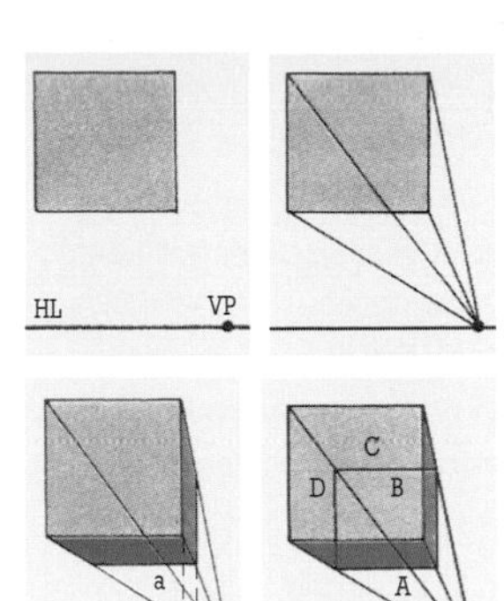

A cube in parallel perspective.

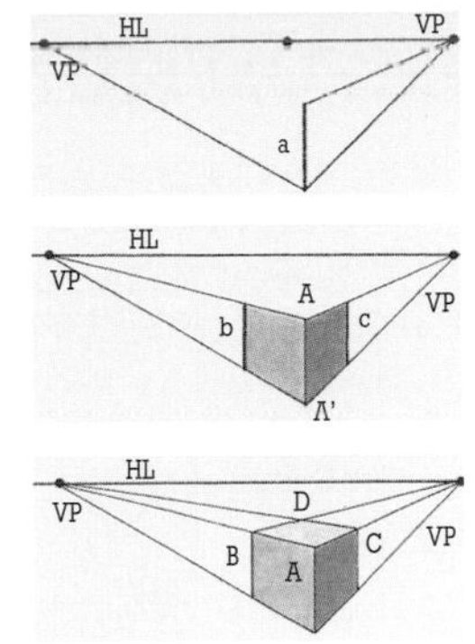

A cube in oblique perspective.

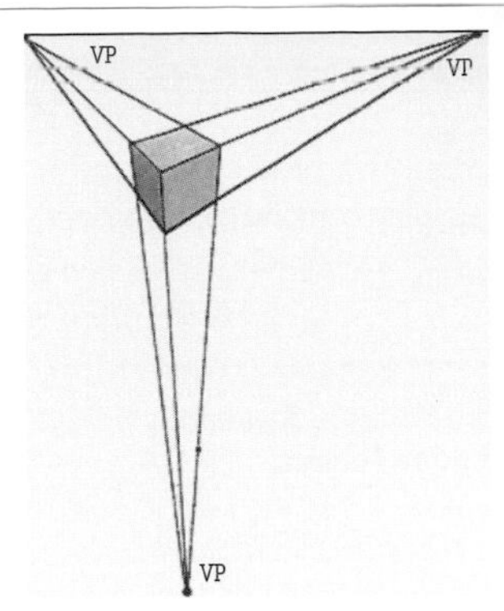

A cube in aerial perspective.

the expression of volume is more complete. In this case, only the vertical lines are parallel.

Using aerial perspective a very precise representation can be achieved, especially if it is not exaggerated. With this type of perspective there are three vanishing points.

In general, the problems of drawing that usually occur in landscapes with water are resolved using parallel or oblique perspectives. Aerial perspectives are particularly used for high horizons, with the presence of boats and buildings.

Differentiating Between Planes

The compositional sketch shows the distribution in a single plane, that of the support, of the different chromatic volumes. In order to differentiate between planes, different tones and values are used, with depth being given deeper value.

To differentiate between the planes, it is necessary to imagine vertical planes that are all parallel to one another, standing perpendicularly to the artist's sight line.

The planes that are closest to the artist will generally form part of the foreground. Here all the elements have greater contrast and definition.

In the background, at infinity, there are a series of elements. These usually include the line of the horizon (which is generally visible) and the whole of the sky.

In the mid-ground are located elements that are going to be depicted with somewhat less detail and lesser contrast.

MORE ON THIS SUBJECT
• Composition **p. 12**
• Depth and Tonal Values **p. 14**

Increasing Depth

In summary, it is useful to be able to differentiate the foreground not only from the background, but also from the mid-ground. Any object or element that appears right in the foreground emphasizes the depth of the painting. Many painters create compositions containing important elements that fill most of the painting's space, drawing the attention of the viewer. But including some detail in the foreground (such as a branch or a bush) allows the artist to emphasize and elaborate on these elements, using strong contrasts in nuance and color. This further heightens the illusion of space by creating contrast with details of more distant elements.

Increasing the depth using elements in the foreground.

Differentiating between planes. Olmedo, Boats. *Graphic representation. Planes: 1) foreground, 2) mid-ground, 3) background.*

COMPOSITION

One of the secrets of converting a theme into an attractive composition lies in establishing an ideal compositional balance between all its elements. It is helpful to create compositional sketches rapidly, trying out different possibilities as you resolve the question of how much weight ought to be given to each element within the whole.

The Focus of Attention. Golden Section

In a well-resolved work there is a point or a zone of the painting that naturally draws the viewer's eye. This is the center of interest. When the composition is too scattered the eye tends to wander, weakening the picture's impact.

On any support, whatever its size, it is possible to establish what in pictorial terms is known as the golden section. Although this is an old theory, it generally gives good results. By the first century BC the ideal distribution of chromatic volumes had already been established by the Greeks. The rule of the golden section is as follows: "an area divided into unequal parts will be aesthetically pleasing if there is the same ratio between the smaller part and the larger part as between the larger part and the whole." In practice, to find the golden section of a painting, you multiply its length by 0.618 to obtain its longitudinal division, and multiply its height by the same value to find the transversal measurement. As the diagram shows, there are four golden points which mark the golden section. If the center of interest is located in this section, with the correct chromatic composition, the painting will be balanced.

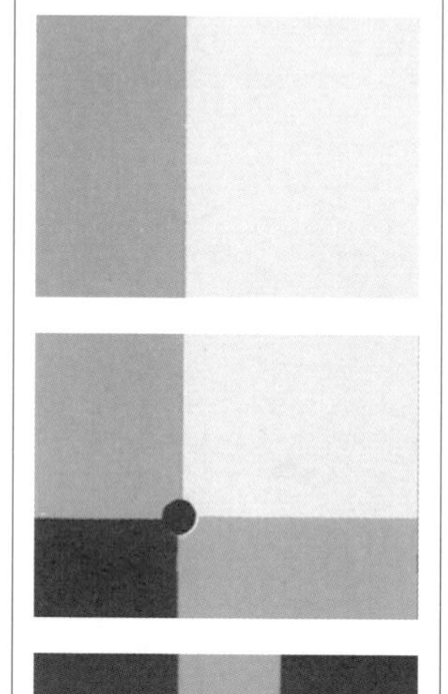

The golden section. How to develop it.

Paul Cézanne, The Gulf of Marseille.

Using the Golden Section

In a work that is correctly composed the viewer's gaze will always be directed towards the center of interest. For example, a work in which a very pleasing visual balance has been achieved is *The Gulf of Marseille* by Cézanne, in which the horizon is situated at the height of the golden division of the painting. Either intuitively or consciously, many painters tend to make use of and apply the theory of the golden section, as it makes for strong and attractive composition.

In example 1, the painter has situated the horizon line in the upper part of the painting, achieving an interesting composition, playing the details of shore and boats against the large, important, reflective expanse of the sea. See example 2.

1. A high or low horizon modifies the result.

Types of Composition

The different types of composition correspond to simple

MORE ON THIS SUBJECT
- The Horizon and Planes **p. 10**
- Depth and Tonal Variation **p. 14**

2. A wedge-shaped foreground, with elements that introduce variations. Winslow Homer, Shore at Bermuda.

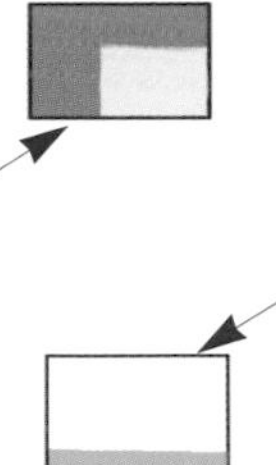

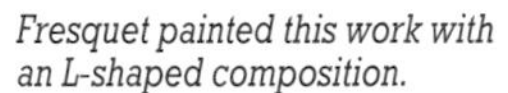

Fresquet painted this work with an L-shaped composition.

Vicenç Ballestar, The Medas Islands, *is an example of lineal composition.*

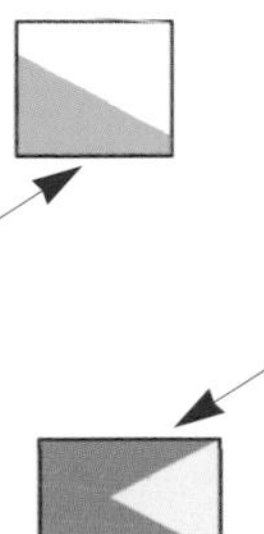

Manuel Plana, Boats on the shore. *This is a diagonal composition.*

Josep Martínez Lozano shows an example of a wedge-shaped composition.

geometrical figures. The most usual are: the composition of a flat line, a diagonal, a circle, a central composition, a triangular one, an L-shape, and so on. What makes a painting with a simple composition special is without doubt the chromatic division of the area of the picture.

Look at a few specific cases: *The Medas Islands,* by Vicenç Ballestar, which is an example of a horizontal composition; *Boats on the shore,* by Manuel Plana, which has a diagonal composition, and the work by Josep Martínez Lozano, which has a wedge-shaped composition.

Chromatic Balance: Balance and Variation

The idea of balance suggests the image of scales. By analogy balance is a question of making chromatic weights counteract with one another, by color and by tone. Sometimes, just by moving the landscape a little and framing it in a different way it is possible to achieve compositions with a more interesting chromatic balance. It is a good idea to establish a composition with a general asymmetrical tendency, by experimenting with variations. In *Cantabrian Harbor* by Josep Martínez Lozano, the chromatic volumes are situated above and below the horizon line, which in turn corresponds to the golden proportion. The diagram illustrates this picture's compositional harmony: the chromatic volumes are balanced.

Seascapes and Lineal Composition

In a seascape the location of the horizon line is very important. There are two options regarding this. The artist can omit picturing it entirely by positioning the painting so that the horizon is higher than the picture plane. Or the artist can break the monotony of the line with elements such as landscape, boats, or buildings, both in the foreground and in subsequent planes.

Josep Martínez Lozano, Cantabrian Harbor.

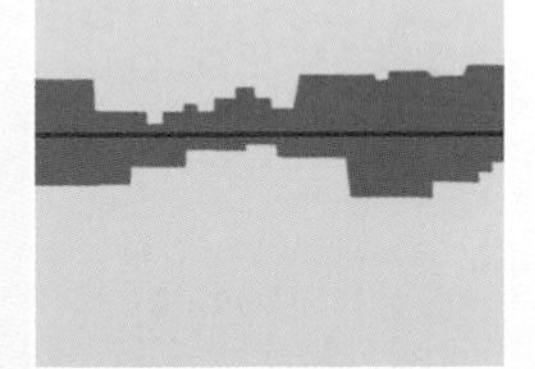

DEPTH AND TONAL VARIATION

Correct tonal values applied to different planes, each depicted with the appropriate degree of precision (more precision for foreground) is what creates the impression of depth on a two-dimensional support. In addition, the color must serve to clarify the shapes and volumes of each element while emphasizing the picture's theme.

Foreground

Everything that is located in the foreground of a painting is what lies closest to the painter. From his or her viewpoint, all the objects in the foreground appear bigger than those that are further away. Only the correct proportions between the elements of each plane (foreground, mid-ground, background) will convey the subject's depth. It is in the foreground that detail and contrast are greatest.

Mid-ground

As the view focuses on elements that are further away, one can see how they diminish in size. The further away they are, the smaller they appear. In this case it is not detail that defines them but more general characteristics. For example, if one visualizes the sea, the foam on the crest of a wave nearby is clearly defined against the rest of the water. However, further away, looking a few meters out to sea, the foam cannot be distinguished and all that one sees is subtle changes in nuance and color. Distance diminishes the details and the contrasts.

Distance

Climatic conditions can exaggerate the effect of the intervening atmosphere. Between the foreground and the horizon, there are many different layers of air. Physically the light and atmosphere interact to produce a series of optical phenomena. Even with a clear horizon, under the best visual conditions, there is a certain loss of color. Detail and contrast diminish progressively towards infinity.

Colors in the distance are usually lacking in intensity. As they grow more pale they are transformed into broken or neutral colors.

In this photograph one can compare the size of the people in the foreground with the people who are swimming in the distance, or those who are on the rocks.

Looking through half-closed eyes can simplify the forms and areas of color.

Distribution of color into simplified zones.

Technique and Detail

The pictorial medium and the techniques used have to express the essence of the landscape with a degree of clarity. A realistic or photo-realistic approach will aim for a greater (almost photographic) likeness of the subject. However, these are not the only possible interpretations. Somewhat unconventional solutions, including abstracts, convey the essence of the water very effectively by means of entirely creative representations.

Josep Martínez Lozano is a watercolor painter who uses colors in a very personal way. His style is free and creative, and based on long experience, making his paintings vivid and splendid. Whatever the subject of the artist's interpretation, there is always a sense that the land or seascape is organized into separate vertical planes or visual fields. His compositions create spaces in which one plane is clearly related to the next. In the work of Armand Domènech one can also observe the clearly organized vertical planes of foreground, mid-ground, distance, in which the three yachts are represented.

MORE ON THIS SUBJECT

- Horizons and Planes **p. 10**
- Composition **p. 12**

Josep Martínez Lozano, Northern village, *watercolor.*

How to Increase the Impression of Depth

Any element located in the foreground has the potential to create the sensation of close proximity. By painting such elements with a high degree of contrast, a strong intensity of color and careful attention to texture and detail, the difference between foreground and mid-ground is greatly heightened.

Armand Domènech composes this seascape with elements in the foreground that make the subject more dynamic.

Division of a picture into planes. A) Three yachts represented graphically. B) Three vertical planes. One can distinguish between the immediate foreground, the mid-ground, and the background.

A

B

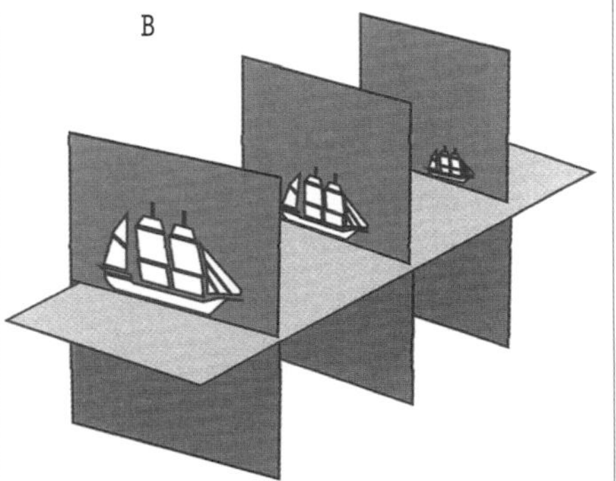

REFLECTIONS

Any water surface, large or small, will create the physical phenomenon of reflected light or images. For reflections to occur, there must be a light source and an element or elements that produce the reflected image. It takes careful observation and analysis to discern the colors of light, objects, the elements of landscape, and their reflections.

Natural Sources of Light: The Sun and the Moon

The sun and the moon are the two most important sources of natural light for the painter. In landscape works that feature a wide expanse of water, the light coming from either the sun or the moon has a great impact on the compositional and chromatic balance. If either of these two bodies appears in the painting, its image will probably be reflected on the surface of the water.

A beautiful and dramatic sunset, its luminous path reflected on the waters of the sea.

Reflections of Light

The artist sees reflections on an expanse of water when located at an angle that receives the rays of light reflected off its surface.

For example, on a very luminous surface the moon will reflect a path running from the horizon, dividing the point of view, towards the viewer. This large-scale reflection of the light of the moon, either on the sea, a reservoir or a lake, will be of such dimensions that its effect on the chromatic balance and the compositional layout must be taken into account. Sometimes changing from a high horizon to a low one or vice versa will create a more attractive landscape, as it will modify the particular chromatic weight of the reflection.

The midday sun, on the other hand, produces a more generalized dominant color on the subject by means of an interactive series of reflections. There are many points of brightness and reflection. Whereas with a sunset, a path of light like the moon's can once more be seen.

A Reflected Image

Any sufficiently illuminated object located directly above a wide expanse of water will produce an image reflected upon it. Depending on whether there is a high or low horizon, the dimensions of the reflected image will vary widely. If the surface of the water is not calm, the reflected image will be fractured and distorted. In contrast, on calm water, there is little distortion.

Distorted or not, the reflected image will always be inverted. It will not be the same size as the object that produced it; it will be the same width, but not the same height.

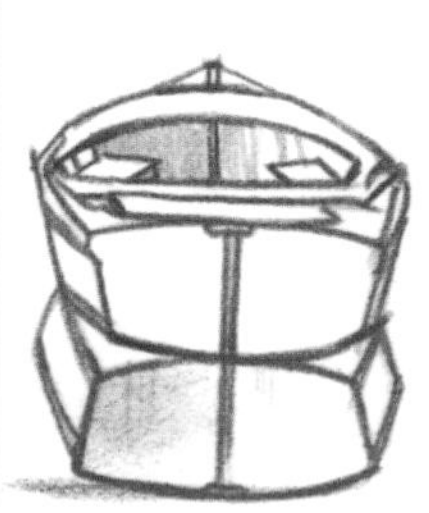

Graphic representation of a boat and its reflection on quiet waters.

Distortions in the reflected images. Any current that causes waves distorts the reflected image.

Perspective and the Reflected Image

Reflected images have the same vanishing point as the figures that cause them, but the lines of the reflection converge on the vanishing point at a smaller angle than the lines of the original form. The reflected image will be vertically shortened.

Although the reflections might be distorted by movement of the water, viewing them through half-closed eyes will confirm that the general chromatic area conforms to the above principle.

In the illustration below one can see the direction of the reflected image of the masts. The movement of the water can cause distortion, but if the boat is upright, the line of the image will follow the vertical of the mast. The complete image of boats and the building is reflected on the water.

The reflected forms have the same vanishing point as the original forms.

MORE ON THIS SUBJECT

- Reflections on the Sea: Light **p. 58**
- The Sea: Reflected Images **p. 60**
- Reflections on the Sea, in the Foreground **p. 62**
- Reservoirs and Lakes **p. 64**
- Reflections in Rivers **p. 68**

Chromatic Weight and the Reflected Image

In this lineal composition (above) reflections are created that reproduce the real image with great accuracy. One can see the dimensions of the image inverted. These allow the artist to give weight to the importance of the reflection of the sky, under this specific type of daylight. The same is true of these examples of seascapes by Josep Martínez Lozano.

In each of these works the chromatic weight of the reflections of boats on sea is different. In the panoramic view, it carries less importance. As the boats become part of the foreground with their reflections, their chromatic weight increases and should be taken into account as part of the chromatic balance of the whole picture.

THE THEORY OF COLOR

Whatever technique is used, the mix of pigment colors and the optical combinations that they permit involve direct applications of a basic knowledge of color theory. The study and application of complete harmonic ranges allows the artist to increase his or her knowledge of the relationships between colors.

Color Theory

Newton and Young, among others, studied the nature of light. From their research it is possible to extract conclusions that are highly relevant for the artist. Firstly and most importantly, without light there would be no color. Deepening one's understanding of the nature of light brings with it knowledge of color and the reason for the coloration of objects.

The composition of white light is the result of the unity of all the colors of the visual spectrum. These can be reduced to three basic colors: green, red, and dark blue. Light travels in waves which causes a reaction when the retina is exposed to them. Each beam of colored light causes a specific sensation in the human brain by means of the retina, which acts as a receiver. At the same time, objects absorb some colors and reflect others. The diagrams illustrate in a practical way the principle of light absorption and reflection.

Additive synthesis between beams of colored light.

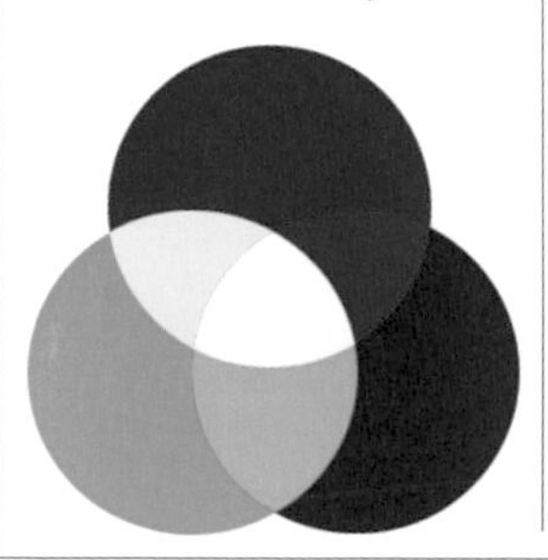

Primary, Secondary, and Tertiary Colors

A painter has to reproduce the colors he or she sees by using paint (colored pigment) and its combinations. If light functions (from a physical point of view) through the additive synthesis of colors, paints work by means of subtractive synthesis. With pigment colors, successive mixtures produce colors that are increasingly dark.

Magenta, yellow, and cyan blue are primary or basic pigment colors. If the three primary colors are mixed in equal proportions, they produce a color that is nearly black. A secondary color is produced by mixing two primary colors in equal proportions. The secondary color green is produced by mixing yellow and blue (cyan) in equal quantities. Similarly, the secondary color red is produced with magenta and yellow in equal proportions. The secondary color blue, which is dark blue, is the result of mixing magenta and cyan in equal quantities. Within this particular theory of color mixing there are three secondary colors: green, red, and dark blue.

Subtractive synthesis for pigment colors.

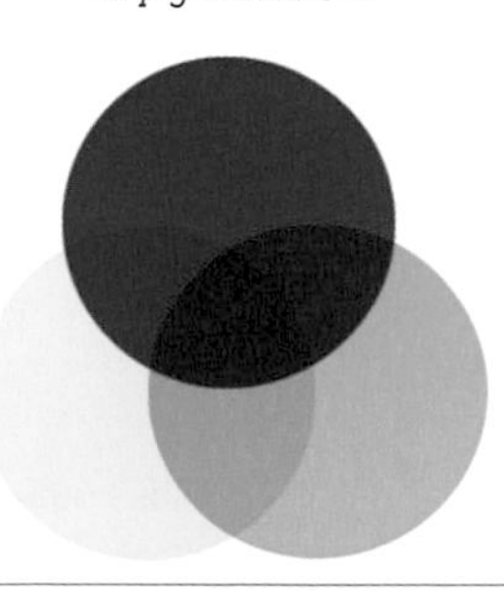

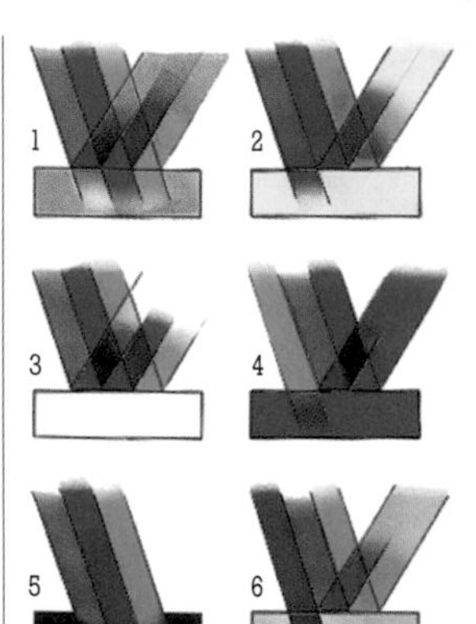

Diagram of the absorption and reflection of light.

A tertiary color is achieved by mixing in equal parts a primary color and a secondary color achieved from the same primary color. There are six tertiary colors. Their composition can be seen in the diagrams which appears on the next page.

Thermal Classification of Colors

Through psychological association, colors are assigned thermal characteristics. They can be warm or cool. This classification is always used as a reference when one is working with non-saturated colors or mixes.

Warm colors produce heat and give the sensation of closeness. Warm colors include rose and red, oranges, yellows, ochres, earth colors, red-purplish blue, and yellowish greens.

Cold colors are greens, greenish blues and blues, and all grays that tend towards one of these colors. On the borderline,

Primary Colors

Primary colors are: yellow, magenta, and cyan.

Secondary Colors

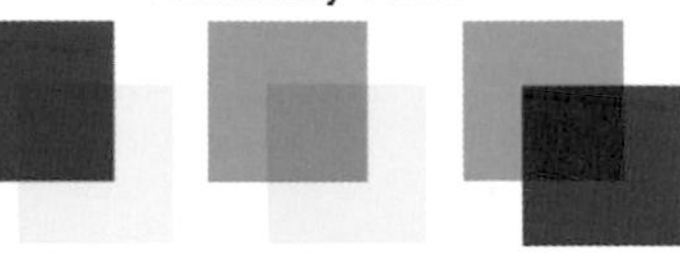

Secondary colors are red, green, and dark blue.

Tertiary Colors

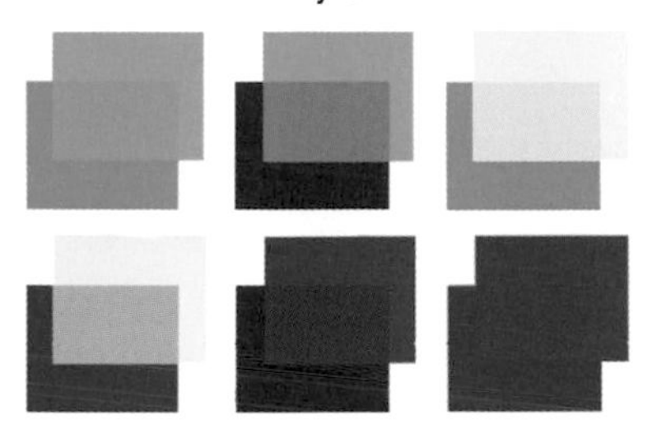

Tertiary colors are: bluish green, violet blue, yellowish green, orange, violet, and crimson red.

but also cold, are certain light greens and purplish blues. Cold colors create a sensation of coldness and distance.

MORE ON THIS SUBJECT

- Oils **p. 26**
- Reflections on the Sea, in the Foreground **p. 62**

Complementary Colors

Within this color system, complementary colors are those colors that produce the maximum contrast when they are juxtaposed. The diagram shows examples. Intense blue and yellow, red and blue (cyan), and green and magenta are complementary colors. Mixing two directly complementary colors in equal proportions gives a very dark gray, almost black, color.

Broken or neutral colors are obtained by mixing two complementary colors in very unequal quantities, possibly also adding white in the case of dark paint. All broken colors have a grayish element, although they can have a marked tendency towards the predominant color of the mixture. A broken color can be either warm or cold.

Yellow and intense dark blue are complementary.

Magenta and green are complementary.

Cyan and red are complementary.

Harmonic Ranges and Fuller Colors

Any ordered succession of colors is called a range. The most simple example is a range created by mixing two colors, merely changing the quantities of each. Color harmony in a painting can be achieved by using various ranges. As a simple example, one range is produced by mixing a color with black and white to achieve a tonal range. Another range uses a color and its complement. Fuller harmonic ranges are made up of warm colors, cold colors, and broken colors.

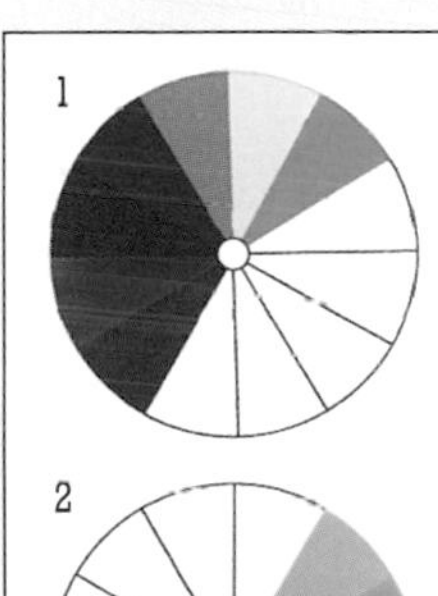

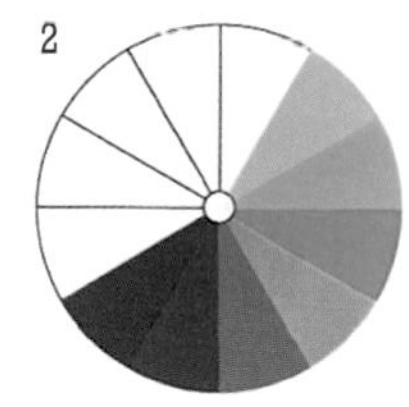

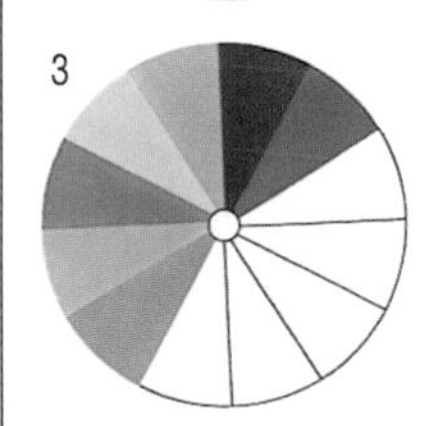

Most common harmonic ranges.
1. *Range of warm colors.*
2. *Range of cold colors.*
3. *Range of broken colors.*

DRAWING MEDIA: INITIAL SKETCHES AND TONAL VALUES

Drawing mediums allow the artist to rapidly produce an initial sketch that explores the compositional design of the work while defining the different planes and the shading or coloring that marks tonal areas. It is well worth the time it takes to do a rapid series of sketches to resolve general problems. The way is then clear for the more elaborate techniques your painting will require.

Charcoal

Charcoal is one of the oldest known existing drawing media. Any subject can be handled quickly and directly using charcoal techniques of line and shading. It is one of the techniques most recommended for beginning large-scale pictures. However one of the limitations of charcoal is its fragility, and work done with charcoal needs to be treated with a fixative to preserve it.

With charcoal one can produce work which is fully resolved in its tonal values.

Although charcoal is a very unstable medium, strong strokes will remain faintly visible after being rubbed over. The whole paper will acquire a generalized shade of gray. The rubbed out charcoal can easily be integrated with whatever media is used to further develop the picture.

Charcoal can be used to make rapid sketches.

A) Charcoal that has not been rubbed.

B) When it is rubbed with a cloth, much of the charcoal disappears.

A general gray appears after rubbing.

Pencil

Lead pencil is really made of graphite, and leaves a more permanent mark than charcoal. It is a tool that is very easy to handle. In general, pencil can be used for very small to much larger works. Pencil's capacity for silvery-gray precision of line and tone makes it a medium that, with the artist's skill, can well represent any subject. Shading and rubbing diagonally, or cross-hatching in successive layers, are some of the techniques pencil offers. A simple variation in the pressure or the direction of the pencil

A tonal range made with a pencil.

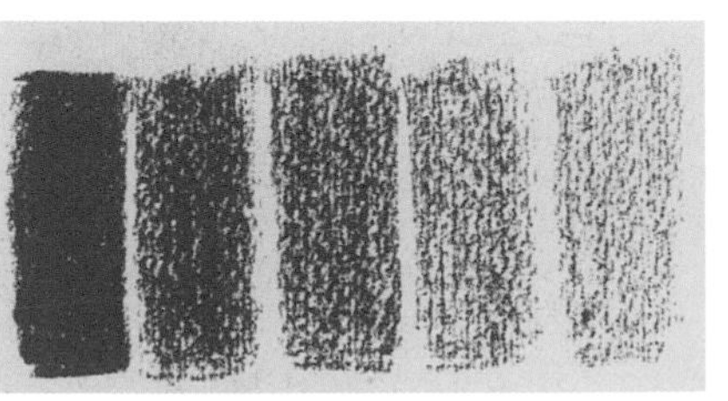

Sanguine. Tonal range.

Initial pencil sketches of different subjects. An autumnal landscape, a winter landscape, and a seascape.

will alter the resulting tone and stroke.

Sanguine

Sanguine is one pigment of the mildly waxed drawing chalk, conte crayon. Conte is also available in black, brown, and white, but sanguine has been used traditionally to work out monochromatic studies, using full tonal values.

Sanguine can be used for creating tonal values of chiaroscuro (light and dark), and to give the suggestion of color. Line and shading techniques for sanguine are very similar to those of charcoal. A stick of sanguine is more hard-wearing than a stick of charcoal. Furthermore, a drawing made with sanguine is far more stable than a drawing made with charcoal.

MORE ON THIS SUBJECT

- Drawing Media: Tonal Values and Color **p. 22**

Sketches and the Finished Work

A sketch or rapid study of a subject can be made using pencil, charcoal, or sanguine, or with any other medium that is fast and easily workable. Monochrome sketches and studies, those in black and white or that use the tonal range of a single color (sepia, sanguine) are a practical tool. These quick sketches resolve questions of framing, composition, definition of planes and general tonal values. The artist can then continue with the coloring and development of the piece. This initial analysis is very useful for addressing large-scale works that require slow development and complex techniques.

Other Media

Although charcoal, pencil, and sanguine are the most usual media for producing sketches, many artists have their own preferences for other media that can be just as practical.

Pastels, for example, are an immediate medium which gives extremely colorful results. Likewise watercolor, although this requires a more elaborate technique of dark on light and the application of wet paint onto a wet, damp, or dry support. A watercolor specialist can obtain spectacular results instantly, in the time it takes to make a quick sketch.

Sketchbook with a rapid watercolor sketch.

DRAWING MEDIA: TONAL VALUES AND COLOR

Pencils, chalks, pastels, oil crayons, and felt-tipped pens can be used to produce full color sketches. Using chiaroscuro, even in a sketch, requires the artist to develop a good knowledge of the practical aspects of color theory as many of its principles are related.

Colored Pencils: Mixtures

Nowadays artists need not hesitate to use colored pencils, since they fully satisfy the prerequisite indispensable for the professional artist of being very resistant to fading. Moreover, the alternatives that are available have multiplied. Water-soluble and turpentine-soluble colored pencils add to the artist's alternatives, in addition to the use of different techniques.

With colored pencils it is possible to create work with very different effects. Using a solvent, a work can display a characteristic smoothness in which the grain of the paper acquires a greater importance. And at the same time, despite being dissolved in water or turpentine (depending on the type of pencils) the shading has an appearance more like that of watercolor, covering the whole surface of the paper.

The most characteristic aspect of this medium is the possibility of mixing colors optically, and this area should be explored by experimentation. In addition to achieving optical mixes with watercolor pencils, or pencils that are soluble in turpentine, it is also possible to obtain homogenous mixes.

Mixes using colored pencils. Primary color gradations. Secondary colors taken from the three primary colors, carmine (magenta), yellow, and blue.

Mixes using colored pencils.

Warm range.

Cold range.

Broken or neutral range.

A seascape made with colored pencils.

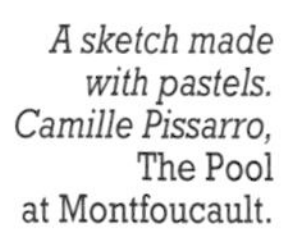

A sketch made with pastels. Camille Pissarro, The Pool at Montfoucault.

Chalks, Pastels, and Conte Crayons: Line, Coloring, Gradation, and Blending

The surfaces of works executed with chalks and pastels are very delicate. The capacity for blending is the most valuable characteristic of these dry media for working color and achieving darks and lights. However the great problem with works in pastel is how to preserve them, as they are quite fragile.

Line and tone techniques are reinforced by the possibilities for blending and shading, both of which are characteristics of chalks and pastels, allowing the artist to create extremely realistic works with these media.

Thinning conte crayon with turpentine.

Conte crayons and oil pastels are grease-based media. Working with them requires different techniques. With conte crayons, the dry drawing-painting can be enriched by using turpentine washes. Oil pastels, which cannot be diffused as can chalk, can be rubbed softly with turpentine to dilute the strokes, so they can be blended. For further information on the possibilities of this medium, see the chapter on oils.

MORE ON THIS SUBJECT

• Drawing Media: Initial Sketches and Tonal Values **p. 20**

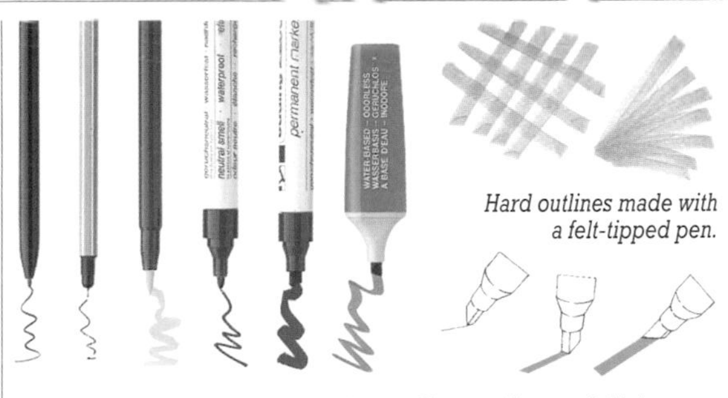

Hard outlines made with a felt-tipped pen.

Lines made with felt-tipped pens. Fine, medium, and thick.

Felt-tipped Pens

Felt-tipped pens are a drawing medium based, above all, on lines and cross-hatching. When they are used on a dry support, their most characteristic trait is the clarity of outline they produce. It is a medium in which the mixture of colors allows for a very lively interaction.

Felt-tipped pens that are soluble in water or alcohol have the capacity for more variation than merely line. Although this medium is known primarily for its application in the context of schools and advertising, the pictorial possibilities of markers are endless. Felt-tipped pens that are soluble in water or alcohol can achieve much plasticity of tone and gradation.

Superimposing tones.

Superimposing colors.

Sketches for a Masterpiece Using a Drawing Medium

To execute *Boats at Sainte Maries* (see page 7) van Gogh made two preliminary sketches. In the first sketch he established the composition and the framing. It has few strokes, but they are sufficient. Notice how the sketches make reference to the colors (below). This was the information the artist required before beginning to paint the work in oil.

Vincent van Gogh, ink sketch for Boats at Sainte Maries.

MIXED TECHNIQUES, 1

For sketching or for creating elaborate, finished work, it is very common to alternate media. Charcoal is often used with white wax crayon, or sanguine with charcoal, or also charcoal with sanguine and white wax crayon. Mixing pastels and wax crayons allows the artist to achieve very singular works, with the addition of the technique of s'graffito.

Historical Mixed Techniques

One of the oldest mixed techniques is the use of black and white media, drawing with black chalk or silver point, with highlights in white conte crayon, on cream-colored paper. The possibilities for darks and lights are enormously increased.

In addition to black and white, sanguine can be used, in what is known as *le dessin à trois couleurs,* to increase the chromatic possibilities available, still using cream paper.

Compatibility

The only condition that the media must fulfill in order to be used in a mixed technique is compatibility. In this sense, all dry media are compatible with one another. As charcoal does not adhere as well as sanguine, it is very different applying first charcoal with sanguine on top than the reverse.

Again, charcoal is very fragile. White conte crayon is more unstable than sanguine. Yet alternating all these media allows the artist to introduce nuances in the way the gradations, blends, and shading behave.

White crayon cannot be used directly on the support if the support is white. However, if white crayon is used in conjunction with charcoal or sanguine, it will be quite visible.

There are media that allow the artist to create work that is line-based but at the same time colored, while offering a wide range of technique possibilities for producing chiaroscuro.

The subject of mixed techniques seems inexhaustible.

In this example, charcoal is used with natural sanguine conte and sepia chalk. There are many possible techniques using colored lines, shading, and blending.

MORE ON THIS SUBJECT
- Mixed Techniques 2, **p. 38**
- Compatibility **p. 40**

Mixtures for dessin à trois couleurs, *charcoal, sanguine, and white wax or conte crayon on cream paper.*

Pastels and charcoal. Blending pastels and charcoal together, can produce beautiful and dramatic effects.

Textures

All mixed techniques increase the possibility of creating interesting textures. Two media as conventional as conte crayon and pastels allow the artist to create major textural works. Pastels can be incorporated easily, since the crayon allows more pigment to adhere to the support without suffering damage when worked over.

S'graffito, which is possible with conte or grease-based crayon, is a technique that permits the artist to create different effects, including recovering the color of the paper by erasing or scraping, producing a flecking effect, narrowing the thickness of the texture and creating different layers.

Imagination and Compatibility

What an artist can achieve by alternating wax or grease-based crayon and pastels depends on his or her capacity for creation.

In expert hands, the possibilities of texture and chromatic warmth can produce surprising artistic effects. In this example, one can see how the pastel adheres to the wax crayon. The layers were applied alternately. The capacity for adhesion was further increased between successive layers by applying fine layers of liquid fixative.

Chromatic possibilities with pastels and wax crayons.

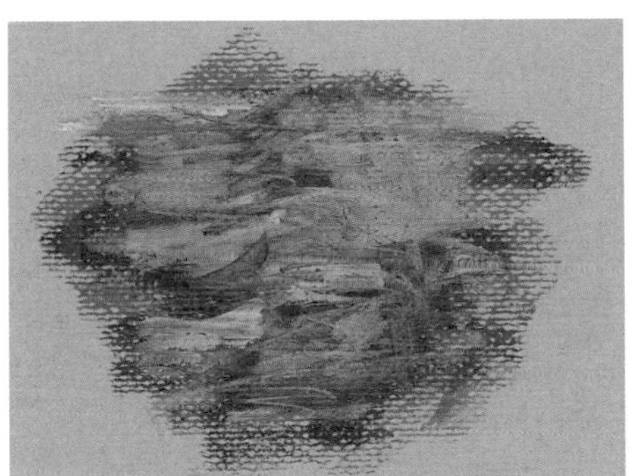

Pastels and oil pastels, with s'graffito. The technique of s'graffito, a combination of working one medium over the other, and scraping to expose layers of color or the paper itself, offers remarkable pictorial possibilities.

A different, completely smooth support, alters the results of the texture.

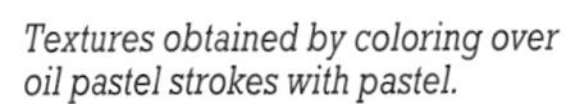

Textures obtained by coloring over oil pastel strokes with pastel.

Oil pastel and pastels, texture. By alternating oil pastel and pastels it is possible to create thick textures. The oil binder of the oil pastel receives the dust from the pastels. Thus the impasto can be built up layer by layer.

OILS

One of the great mediums historically is oil painting. Oils are so versatile that almost any effects of color or texture can be achieved. The fact that one must work layer by layer, and that oil paint is very slow to dry are two major factors that affect the way the medium is used.

Mixing Colors in Oil: Blended Paint and Optical Mixes

Oil paint can be mixed on a palette or directly on the canvas, worked with wet paint on wet paint, or with diluted paint on dry paint. From a base of three basic or primary oil colors, such as rose madder, winsor or thalo blue, and cadmium yellow, plus white, almost all the mixtures necessary to obtain any other color can be created. This exercise requires a certain skill, so it is more usual to use a palette with at least twelve premixed oil colors to ensure a full spectrum.

Colors should be lightened with similar colors from the same color group and darkened progressively with dark colors and "blacks." In the medium of oil colors, blacks (or chromatic grays) can be mixed from rose madder, chromium oxide and burnt umber; or cadmium red, ultramarine blue, and burnt umber.

Mixtures of oil paints. A mixture for a grayish, blue-green. Here ultramarine blue and thalo (or you can substitute winsor) blue have been mixed, with a little ochre to add nuance.

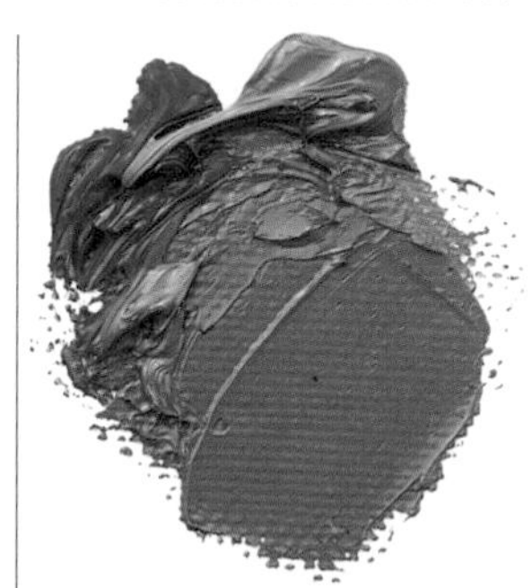

Blended paint. When two or more colors are blended, the same color appears throughout the mixture.

Optical mixture. The mixture is produced in the human retina when viewing the juxtaposed patches of color at a distance.

Oil paint applied transparently. Notice the difference when the color is applied opaquely.

Applying Paint Transparently. Glazes

Oil paint can be used transparently, diluted with a mixture of linseed oil and turpentine. By modifying the density of the paint, you can produce a tonal range of a color, in other words a gradation of transparencies.

When the layer of paint is very fine, its color highly diluted, is applied over another color that is already dry, its film alters the appearance of the color underneath—glazes it. The resulting effect will vary depending on whether the paint beneath is opaque or transparent. In the latter instance, a mixture of color glazes is produced that are similar to superimposed layers of watercolor. Transparent oil paint is very luminescent, and its appearance is oily and shiny if only linseed oil is used.

Mixtures made by superimposing wet-into-dry glazes, which are highly diluted layers of oil paint.

Opaque gradations. Black has been gradated with white.

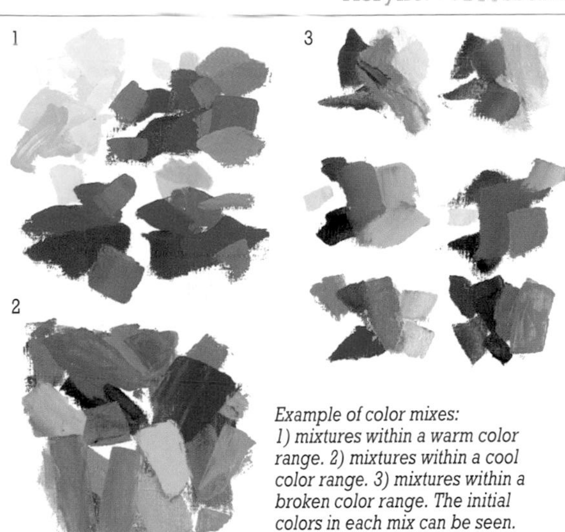

Example of color mixes: 1) mixtures within a warm color range. 2) mixtures within a cool color range. 3) mixtures within a broken color range. The initial colors in each mix can be seen.

Applying Opaque Paint

Oil paint can also be used opaquely. If the density of the paint is left practically the same as when it comes out of the tube, it is so opaque that it can be used to correct mistakes by painting it directly over dry work, with the result that the layer of paint beneath will be completely covered.

Gradations of Opacity and Mixtures

In order to create ranges of colors, in other words ordered sequences of colors, with opaque paint, it is necessary to apply a minimum of color theory. A tonal range made with white is one of the most simple, and produces an opaque gradation. Ranges using only two colors can be made by mixing the same two colors in different quantities, progressing in even stages from one color to the other.

Any oil color, either pure or mixed, to be used opaquely, can be graded by mixing it with white. Just two colors and white can produce a surprising variety of colors.

With regard to the range of mixing possibilities for oil colors, remember that primary colors will mix secondary colors, and primary and secondary colors will mix tertiary colors. Experimenting with mixing in this way will substantially extend your palette of colors. Color mixing possibilities are limitless and good results take practice and exploration.

MORE ON THIS SUBJECT

• Color Theory p. 18

Harmonic Ranges

Notice the three sets of examples of oil paint mixes. The first consists of warm colors. The second is a selection of cool colors. The third is composed of colors that are neutral, including grays, some of which are very dark. These mixes are broken colors, also known as neutral colors or chromatic browns and grays.

In the color mixtures of example 1, warm colors have mostly been used with a touch of cool to add nuance. In example 2, combinations of cool colors have been used, as well as a little warm color to give nuance.

The mixtures of broken colors in example 3 are obtained by mixing complementary colors in unequal quantities, with or without the addition of white.

Craft

Oil paint can be worked on in layers (greasy layers on thinned layers), as a transparent or an opaque paint, with impasto techniques, using s'graffito or scraping, blending, shading, smearing and rubbing, and also in the direct method of painting known as *alla prima*. With oil painting it is useful to plan the sequence of techniques and processes to follow. It is a help, and takes some experience, to be able to assess when an oil painting can be worked wet-into-wet, or into dry, in order to achieve the desired results. For example, for clouds that must be painted with mixes of color, the paint should be applied wet-into-wet.

Clouds in oil.

ACRYLIC: PROCESSES AND AUXILIARY MATERIAL

The medium of acrylic is another of the great pictorial processes. Once it is dry, acrylic paint is permanent, thanks to its polymerization, and it is very resistant to light and fading. It is also very versatile. While oil paint dries very slowly, acrylic is a medium that dries very quickly.

A Water-soluble Medium

The acrylic polymer emulsion can be diluted in water while it is damp. Later, as it polymerizes as it dries, the paint becomes a resistant, flexible layer that does not yellow, although it does tend to contract in thickness.

Acrylic colors.

Acrylic Colors

For acrylic paints, permanent pigments are used. The most usual ones are based on quinacridone, and pthyalocyanine. Here is a possible palette for painting with acrylic:

Cadmium red, quinacridone red, naphtol crimson, red iron oxide, azo yellow, yellow ochre, raw sienna, raw umber and burnt umber, pthyalocyanine blue, ultramarine blue, dioxanine purple, phtalocyanine green, chromium oxide green, titanium white, ivory black.

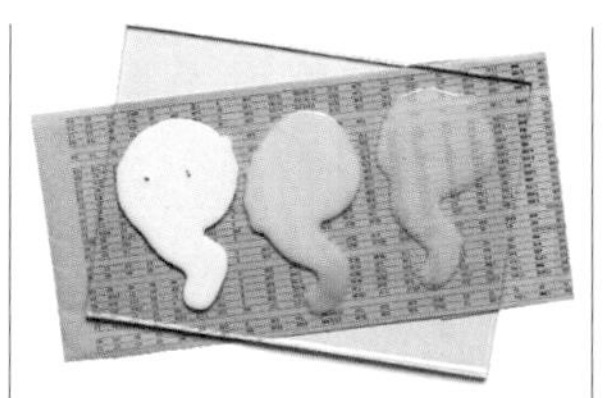

An example of three types of media suitable for acrylic paint, from very opaque to very transparent.

Acrylic Medium

The medium most suitable for thinning acrylic can be gloss or matte. There is also gel medium that can extend the color or make it more transparent. All of these are compatible with acrylic painting.

Bearing in mind the great speed with which acrylic paint dries, it is useful to use a retarder, which is usually available as a gel or a liquid.

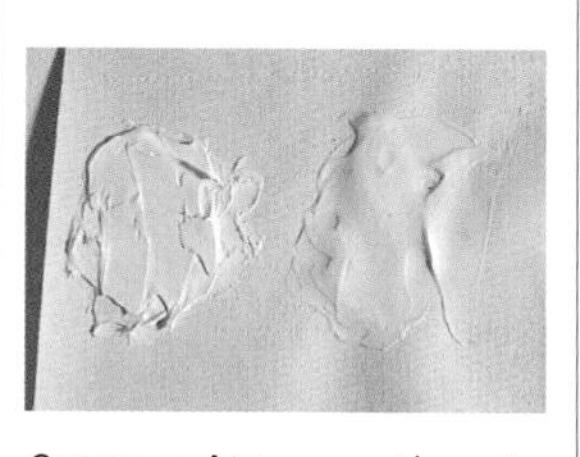

Opaque and transparent impasto.

Acrylic Paints with Transparent Techniques

Unlike watercolors, in which even a dry layer of paint is always soluble in water, once a transparent layer of acrylic paint is dry it does not deteriorate and alter when it is painted over.

There is a whole series of techniques for smoothing edges and softening lines which cannot be used when the acrylic paint is dry.

Transparent Impasto and Opaque Impasto

Thanks to gel it is possible to produce transparent impasto effects. This is when the artist applies paint thickly but transparently by mixing a quantity of

Mixed with gel (thick paste) acrylic paint acquires the texture of oil paint.

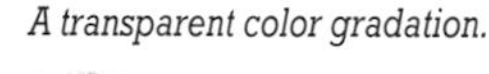

A transparent color gradation.

gel with pigment. The possibilities for creating effects are multiplied with transparent impasto, semi-transparent or opaque impasto, glazes against a white background or color that is smooth or of varying texture.

Opaque acrylic colors can be mixed in a similar way to oils, except it is advisable to add a retarder to mixtures in order to be able to work without a time pressure.

Acrylic paint can be used opaquely, like gouache, although the two media do not flow in the same way, as gouache has a flat, matte finish. Given the resistance of acrylic (when it dries it is insoluble) it is advisable to use it when its specific qualities are desired.

An example of acrylic paint over-painted with oil paint, wet-into-dry. (Acrylic should not be painted over oil, however, as it will crack.)

Combining Techniques with Acrylic

As with any other medium discovering the potential of the medium of acrylic will take exploration and practice. But the range of its effects can be very exciting.

Techniques for Spreading the Color and for S'graffito

In the form that acrylic paint comes out of the tube, it can be spread over the painting surface using a brush or any instrument that has a straight, flat edge. A layer of paint that has just been laid and that is still damp can be easily scraped. If the dry surface of the paint is scraped with a pointed object, a screwdriver or a knife, it will reveal the color beneath, and leave linear marks as well.

By scraping, it is also possible to mix damp acrylic colors. The drawing can be made by passing a scraper over the damp, painted surface, or even over freely drawn lines which will blend with the other colors.

This type of painting permits the use of stencils and masks of all types, including masking fluid.

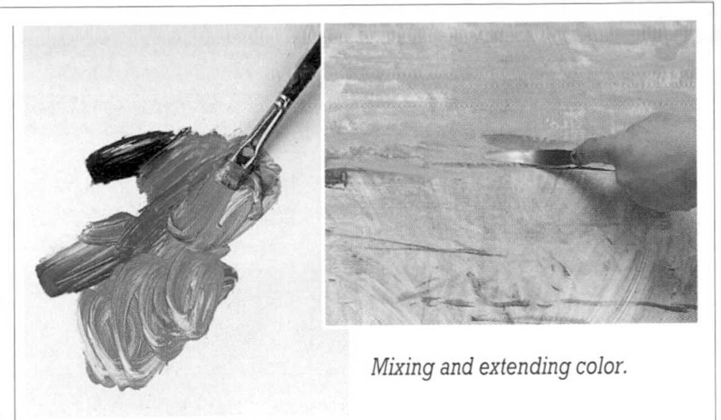

Mixing and extending color.

MORE ON THIS SUBJECT

• Color Theory p. 18

S'graffito on paint: 1) wet acrylic and 2) dry acrylic.

Extruded Acrylic

Another of the possibilities of this type of paint is what is called the extruded technique, squeezing the paint from an icing bag or directly from the tube. Acrylic paint straight from the tube has a tubular appearance. By leaving it to dry you can create strands that are elastic enough to be further worked once they are dry.

Extruded acrylic. Notice the sections of tubular paint (just as it came out of the tube), now dry. They are elastic enough to be manipulated.

WATERCOLOR: TRANSPARENCIES AND EFFECTS

The most important characteristic of watercolor is its transparent nature, so that dark colors are traditionally applied over light colors.

Painting with Watercolor

Watercolor is based on a transparent technique that is used by superimposing fine washes of transparent colors and using the whiteness of the paper to illuminate the paint. The more layers that are superimposed the more tone and depth the paint acquires. As it darkens it also loses luminosity.

Watercolor is water soluble and can be diluted and modified in many ways by adding varying amounts of water and applying it with paintbrushes, sponges, cloths, blotting paper, tissue paper, cotton sticks, and so on. It can even be modified once dry, by adding or even removing color.

Wash Techniques

A wash is a very fine layer of paint highly diluted in water, which completely or partially covers the surface of the support. Part of the art of watercolor lies in the skill with which the artist applies washes. A wash can be applied homogeneously as a background, covering the whole of the support. This layer of color acts as a base for applying successive washes, layer by layer. Similarly a wash can be graded, from light to dark or from dark to light.

Even or graded washes painted onto a damp support give a very different result than those painted onto a dry support. This is a question that is very important for applying watercolor in any manner. Working wet into wet, or wet on a dry support give two distinctive effects. Whether even in color or graded, washes can be altered by adding water, or removing color.

The use of masking fluid and wax can also produce interesting effects.

Different effects are produced with uniformly-toned washes on dry paper, uniformly-toned washes on damp paper, graded washes on dry paper, and graded washes on damp paper.

A graded wash.

An homogenous or even wash.

An example of wet-into-damp.

An example of wet-on-dry.

Superimposing layers, wet-on-dry.

Superimposing Colors

As a transparent paint, the mixes created by superimposing layers of wet paint onto dry paint can be very clean and pure. The application of dark paint onto light paint will ensure transparency and give the surface a luminous quality.

Working in the layers of color which watercolor paint requires, the painter is directly applying principles of color theory. A blue layer superimposed onto a yellow layer will give a green color. A rose magenta layer superimposed onto a yellow layer will give the color red. A layer of rose magenta superimposed by a layer of blue will give a dark blue color.

Wet-into-Wet Techniques

The illustrations show some of the effects that can be achieved. In the first example, a blue brushstroke has been applied over a yellow background while the yellow was still damp. Another example shows the result of applying patches of various colors separately onto a damp background of orange.

The application of wet-into-wet watercolor produces a filament-like effect. The paint will modify its appearance in one way or another depending on the characteristics of the paper or support.

Mixing colors using this technique can produce beautiful results. Even though the painter might do things deliberately, the results can never be predicted accurately. The produced effects will depend on everything from the amount of paint and the dampness of the paper (whether uniform or otherwise), to the way the support is moved (for example, tilting it so that the paint runs in one direction or another.)

An example of wet-into-wet.

When working in glazes or layers the layer of paint below should be completely dry. The application of the following layer should be made with great delicacy, taking care not to disturb the base layer.

How to remove wet color.

Sponging Out

With the help of a clean, dry paintbrush it is possible to remove some of the paint while it is still wet, leaving a patch of color that is more transparent against the white of the paper.

With the same process it is possible to lift patches of color where liquid paint has collected, which may, in some cases, be undesirable, since pooling produces colors that are darker than the rest of the color area.

MORE ON THIS SUBJECT
- Color Theory **p. 18**
- Watercolor: Techniques **p. 32**

Reducing the Color Once It Is Dry

In the case of a patch of watercolor that is completely dry, it is possible to reduce the intensity of the color by applying water to it. Or, when still wet, water and excess color can be removed using a clean, dry paintbrush.

Other drying materials can also be used, and will produce a variety of effects. A cotton stick will lighten small patches. Drying the whole painting with paper toweling can give a textured, grainy effect.

Reducing the color of dry paint.

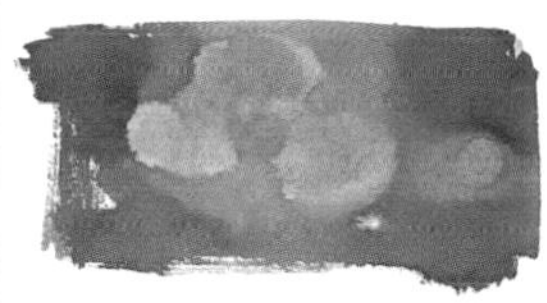

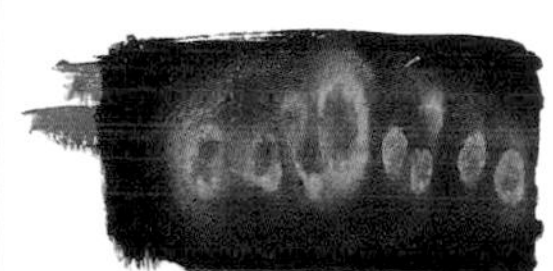

Effects with Salt

Sprinkling some grains of salt onto a damp wash of watercolor paint produces very interesting effects. The salt absorbs the dampness of the paint and crystallizes to create different shapes. The appearance of the surface will be affected by the amount and kind of salt and by the dampness of the paint.

Salt and watercolor. Creative effects.

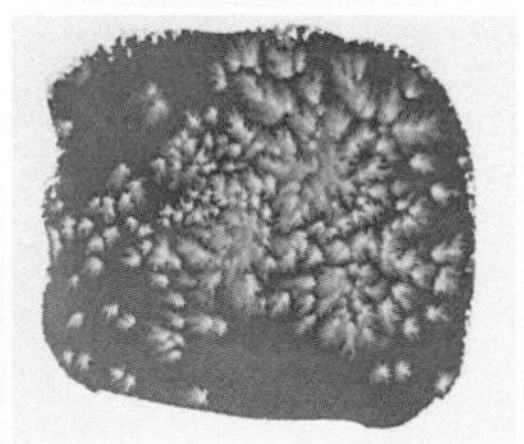

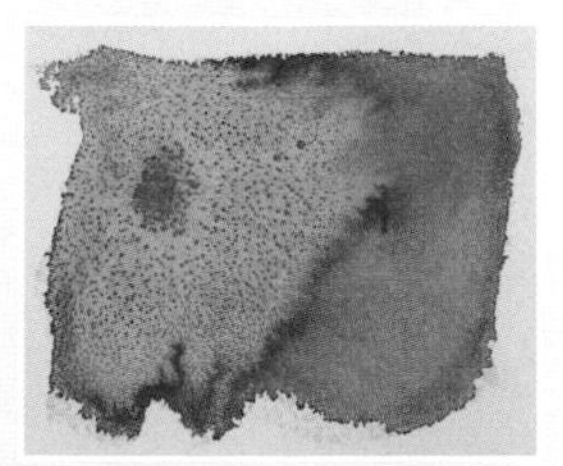

WATERCOLOR: TECHNIQUES

As watercolor is a medium that allows the artist to work with techniques of wet-into-wet, wet-into-damp, and wet-into-dry in any combination, there are ways of working the paint that are very useful for representing water specifically. For example, softening outlines and creating white patches.

Softening Outlines or Overly-Harsh Edges

When the artist works on a dry base, the outlines of the brush-strokes and patches of color tend to be hard-edged. These edges can be softened in various ways. In the example in the illustration, with a clean paintbrush saturated with water, the edge of a patch of dry color that was applied to a wet base has been softened. It is necessary to wait for a few seconds while the water lightens the color. On a damp patch of color, with the same technique the edge can be softened to give a frayed effect.

A softened outline.

Creating White Patches

Given that even dry water-color can be lightened with water, it is alright to let the paint dry before creating white patches. To draw clouds (or the foamy crests of waves) on a background of completely dry blue paint, use a clean paint-brush saturated with water to dampen the areas where you wish to recover the whiteness of the paper. Wait a few seconds and then remove the damp paint with the help of another clean, dry paintbrush, or by using a blotting paper.

If the edges are too harsh, remember to soften them using a wet brush to apply water.

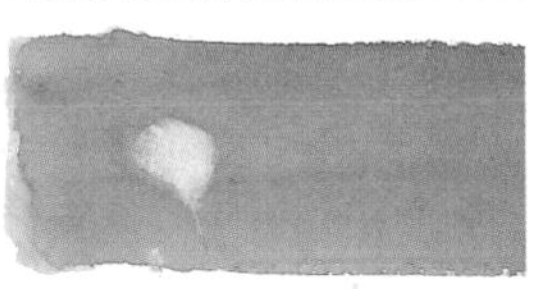

Creating white areas on dry paint.

Scratching Out

It is possible to produce interesting effects of tone and color using the technique of scratching out to open up white areas and to create contrasts in texture.

With the help of a knife or the point of an x-acto tool, and taking care not to damage the surface of the paper too much, it is possible to scratch out the paint of a wash of color that is completely dry. But be sure the paper or support is completely dry. Otherwise the sharp tool might gouge the paper too deeply.

Softening

The delicacy of a picture painted with watercolor depends on the skill the painter develops for softening or avoiding harsh edges where wet paint is applied to paint or paper that is dry.

Wet-on-Dry Technique

Washes of the same color superimposed on one another, applied wet-on-dry, create tonal gradations of this color. This is a property that is often used to create monochromatic paintings.

In a painting in which various colors are used, they can be progressively darkened with superimposed wet-on-dry washes of the respective colors.

As the example shows, it is possible to produce various tones of the same color. When using this technique start with pale tones of the chosen color,

Scratching out or s'graffito to produce white patches.

Wet-on-dry tonal gradations of a single color.

using it in a very diluted state. After letting the first layer dry, subsequent washes are applied with more intense color. The lightest areas indicate the most illuminated areas of the subject, and so on. With this technique, the contrasts progress very quickly with each new application of washes.

Superimposing Washes of Different Colors

In the previous section it was shown that the tone of a color can be intensified by superimposing a layer of the same color. If different colors are used, color areas will be darkened or altered.

MORE ON THIS SUBJECT

- Color Theory **p. 18**
- Watercolor: Transparencies and Effects **p. 30**

When complementary colors are superimposed, the darkening effect is more marked, even with very fine layers. This technique can be used to create strong contrasts or blocks of intense color.

Watercolor is one of the most difficult media to handle with skill. It is always better to use fewer stages of work, as the end result can be dull if the painting is manipulated too much.

Modifying a color by superimposing wet layers onto dry paint.

To Superimpose or to Mix Color?

Watercolor techniques can easily be used in conjunction with one another. The artist can alternate a succession of washes, gradated or otherwise, paint wet-into-wet or wet-on-dry, wash out white patches, or scratch out color. Mixes can be created on the support, painting wet-into-wet or wet-into-damp. Watercolors can also be mixed on the palette or in pots. Any mixture can be used to paint wet-into-wet or wet-on-dry. It is important to work carefully, thoughtfully assessing the development of your piece, as watercolor does not allow for too many corrections.

Alternating techniques. Josep Martínez Lozano, Port.

GOUACHE: A WATER-BASED MEDIUM

Gouache is a water-based medium that comes in the two most convenient forms: in tubes and in cakes. Its purposes stretch from large jars used by learning youngsters to high quality products for commercial and fine artists. The most important characteristic of gouache paint is its ability to cover surfaces and its opacity.

Opacity and Transparency

Gouache paint is a medium with a lot of body. It contains more glycerin than do watercolors, and a thickening agent such as barium sulfate or chalk, which allows the pigment to absorb more water. With gouache it is possible to apply washes that are flat, opaque, and even. Such is its opacity that when it is applied in quantity it can completely cover the color of a support or any underlying layer of paint.

Watercolor needs a white, or at least light colored, support so that the paint will appear transparent and luminous. This is not the case with gouache, which can completely cover a support, however dark it is.

Gouache covers the color of the support entirely.

Applying an opaque color (cadmium red) completely covers the vermilion red.

When used opaquely, gouache can be applied either as dark on light, or as light on dark.

At the same time, when highly diluted with water, gouache can be used in semi-transparent coats. However, highly diluted gouache has a less luminescent appearance than watercolor.

Alternating Techniques

When used at its full opacity, gouache can color spaces evenly and smoothly. It is a medium that is often used by artists who work in the fields of design and advertising.

When diluted with sufficient water gouache permits the use of very fine, semi-transparent washes. These coats (again, which are less luminescent than those of watercolor) can be superimposed on opaque layers. As they are not completely opaque, the mixtures that are obtained allow the artist to build contrast more

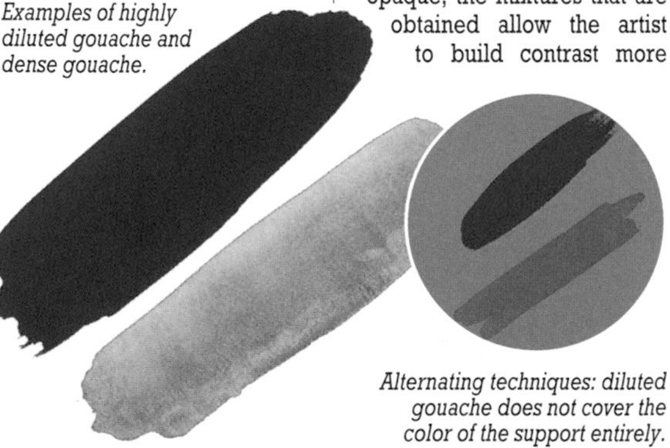

Examples of highly diluted gouache and dense gouache.

Alternating techniques: diluted gouache does not cover the color of the support entirely.

The opaque nature of gouache. It is possible to paint the background first. The work is completed by painting on top of this base. Successive layers and brushstrokes cover the paint beneath.

MORE ON THIS SUBJECT

• Color Theory **p. 18**

slowly. The examples of different patches of color illustrate some of the effects that can be achieved by altering the density of the gouache.

Gouache is a very easy medium to handle. It is very pure in color. It can represent subjects loosely or in great detail. This painting shows the work of colorist Miquel Ferrón.

Watercolor and Gouache

As they are both water-based media, watercolor and gouache can be combined, and are entirely compatible. Watercolor requires a white support, or one that is more or less light, in order to achieve transparencies. It is possible because of its opacity to paint a colored support with white gouache, then use this primed support as a base for painting with watercolor.

The visual effects of combining all the possibilities of both media are endless. They can be used as transparent, semi-opaque and opaque layers. The entire range of watercolor techniques can be combined successfully with the opaque qualities of gouache.

Watercolor and gouache: alternating techniques. There is a difference between the transparency of watercolor and the semi-opacity of highly diluted gouache.

INK: NIBS, REEDS, AND BRUSHES

In addition to traditional black or sepia inks, nowadays there are various colored inks available. High quality products can be obtained from suppliers, including strong red, blue, or green inks, and yellow inks which are more delicate. It is very important to understand the qualities of the ink one is working with. Permanent inks cannot be corrected once they have dried. The beauty of ink colors is their brilliance, their intensity, and their transparency.

Pen and Ink

Using a nib, ink can be applied using techniques of line, shade, and cross-hatching. The artist uses the shaft of the pen to control the direction and precision of the line it produces. In this way the artist can produce work that uses the beauty of line as well as tones and shading, achieved by density of line and by cross-hatching.

With a nib it is possible to produce all types of strokes, as long as the nib is flexible enough to register varying amounts of pressure.

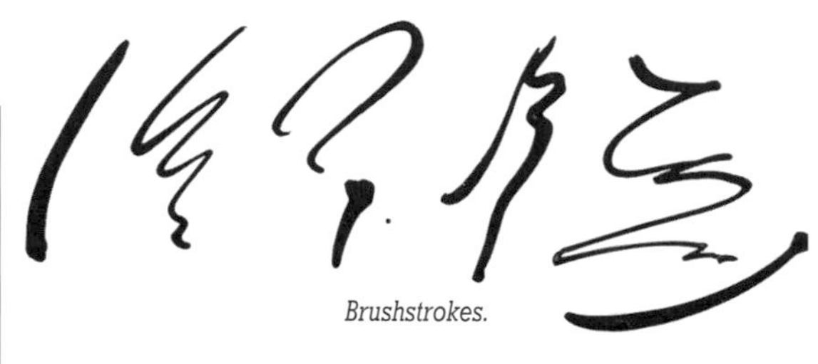

Brushstrokes.

Cross-hatching with a Nib

There are a number of ways to create tones and shades with ink: shading in one direction, diagonally, across-wise or in spirals. Gradations can also be achieved using zig-zags or pointillism (stippling). See the small selection of examples. The possible methods for producing tones and shades are many.

Differentiating Planes

By creating layers of lines, it is possible to construct a tonal range. Differences of tone will delineate and define different planes.

One can combine stroke techniques in the same drawing, producing varying optical effects and creating differentiations between planes. A broad spectrum of tonal gradations can be achieved by using a single method, and many more gradations and effects can be achieved when combining several methods.

A repeated stroke in one, or changing direction can create visual rhythms and increase the effect of depth in a work produced with pen and ink.

See the examples of stippled and zig-zag hatching. To control contrasts, the stippled, pointillist method is more precise and delicate than hatching, which tends to be bolder and darker.

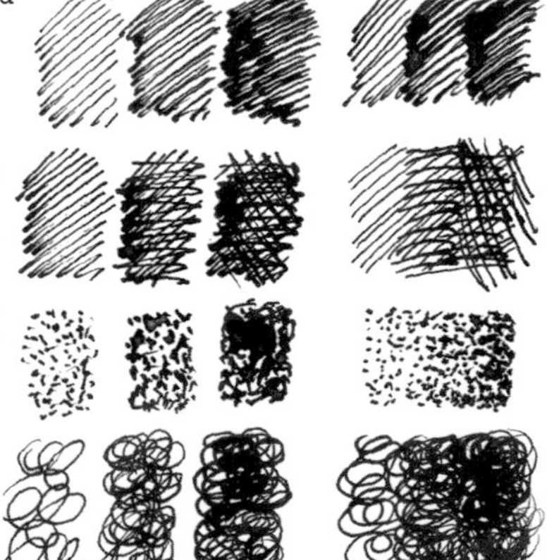

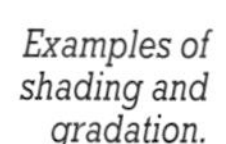

Examples of shading and gradation.

Reed Pen and Ink

The reed pen is an instrument that can be purchased, though it can also be made easily using a reed. To do this, cut the reed diagonally, then make a lengthwise cut, dividing the two halves and another in the center so that it will hold more ink. Several kinds of reeds can be found in marshy areas, or obtained at a florist.

When drawing with a reed, there will tend to be more gray colors, both light and medium in tone. This effect can be empha-

The use of shading to differentiate between planes.

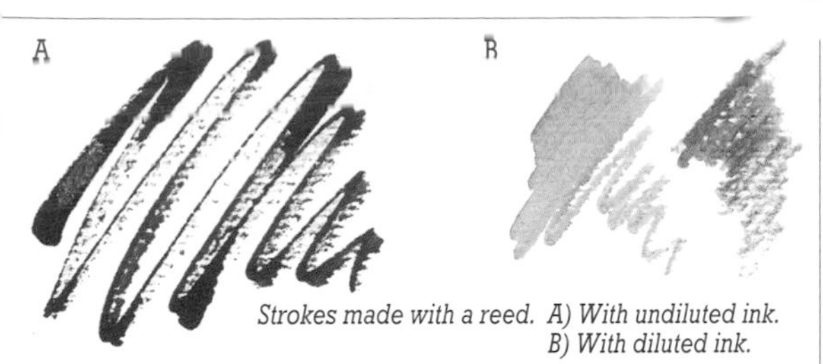

Strokes made with a reed. A) With undiluted ink. B) With diluted ink.

sized further by using ink diluted with water. Works done with a reed pen are looser and have a freer, washier quality than works done with a nib.

Even the most malleable nib is less flexible than a reed pen. The two implements move across the paper in very different ways.

Notice the different qualities obtained by using diluted ink and ink to which water has been added.

Brush and Ink

With a brush one can make straight brushstrokes of varying thickness, as well as free, loose strokes. It is possible to play

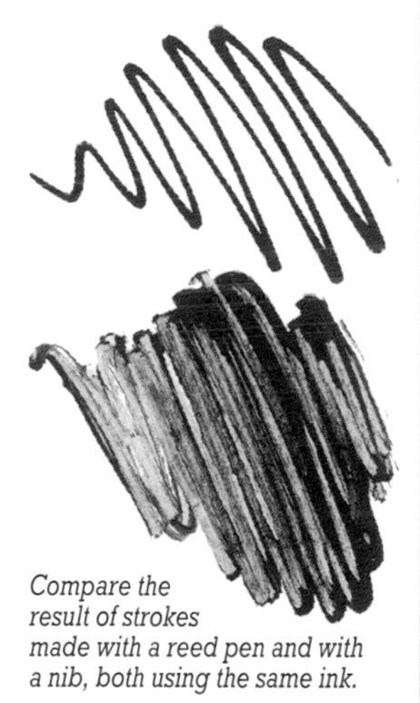

Compare the result of strokes made with a reed pen and with a nib, both using the same ink.

with the dimensions of the paintbrush and the marks it will make using different pressures exerted on its tip or along the entire length of its bristles. Practice is important, and with experience the artist will learn to keep the wrist flexible which will add to the beauty of the stroke.

Monochromatic Washes and Tonal Values

When diluted with distilled water, ink becomes very transparent. A monochromatic wash can create a tonal range. Diluting the ink diminishes the intensity of its color and makes it possible to create color gradations.

It is also possible to achieve a tonal range by painting multiple layers, wet-into-dry, using the same color density in each layer. The superimposition of each successive layer will make the drawing darker. By using this technique it is possible to maintain the definition of all the outlines.

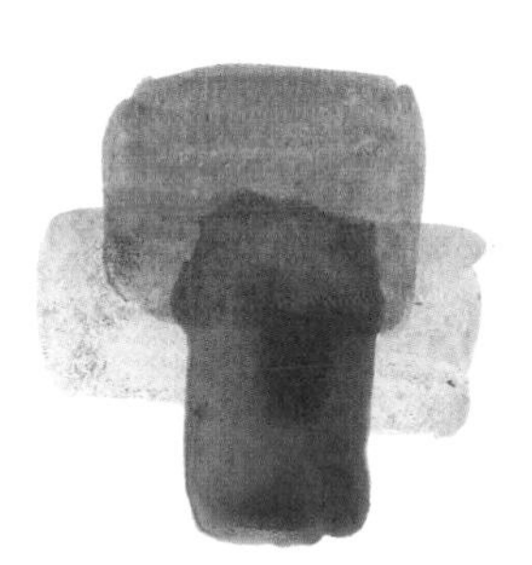

A monochromatic wash. A tonal range.

Colored Inks

In addition to the traditional inks, black and sepia, inks in other colors are also available. Although there is not a wide color range, it can be extended by superimposing layers or by mixing colors, as shown in the examples. Notice how beautiful ink colors are. They are intense, brilliant, and transparent.

Colored inks.

MORE ON THIS SUBJECT

• Color Theory **p. 18**

Alternating Processes: Colors and Lines

Making a drawing in ink using a pen and brush, using dilution or introducing various colors, enormously increases the artistic possibilities of the work. At the same time it is difficult to make corrections with ink (since this involves damaging the support by scratching). But its beauty is worth the exploration and experimentation it takes to master the medium.

Alternating color (brush) and line (nib).

MIXED TECHNIQUES, 2

The following are a sample of the most common processes that fall under the heading of mixed techniques. They are all simple to employ and create surprising artistic effects, including the enormous possibilities regarding texture and finish.

Wet and Dry Media: Watercolor, Gouache, and Pastel

Watercolor and gouache are water-based media. Again, the difference between them is watercolor's transparency as compared with the opacity of gouache. But remember, the latter can be used both opaquely and highly diluted (see the section "watercolor and gouache" on page 35).

A surface colored with pastel to which a few brushstrokes of water have been added.

Pastels are a dry medium that can also be used with water. The strength of this mixed technique lies in the contrasts that can be obtained between the vibrant dusty appearance of the pigment in dust form and the watery transparency of watercolor. When pastels are damp-ened with water their appear-ance changes entirely. If you paint over a pastel work (which is always very delicate) with gouache, you will achieve an impasto (layered or textured finish) which can be both expressive and dramatic. The two media together create interesting contrasts.

Oily Media: Oil Crayons or Sticks and Spirit-diluted Oils

Oil crayons use various oily binders for the pigment. These

A

B

C

A) The support (canvas) is colored with oil crayon and painted over with oil.
B) Thinning the oil crayon and paint with thinner.
C) Wax has been melted and then applied, and then painted over with oil paint.

Examples showing the different characteristics of alternating watercolor, gouache, and pastel.

1. *A) The support has been colored with sky blue pastel.*
   B) A brushstroke of watercolor is added, and another of very thick gouache.
2. *A) Here two brushstrokes, one of watercolor and the other of gouache, are applied to the blank support.*
   B) On top of the dried brushstrokes, sky blue pastel is added.

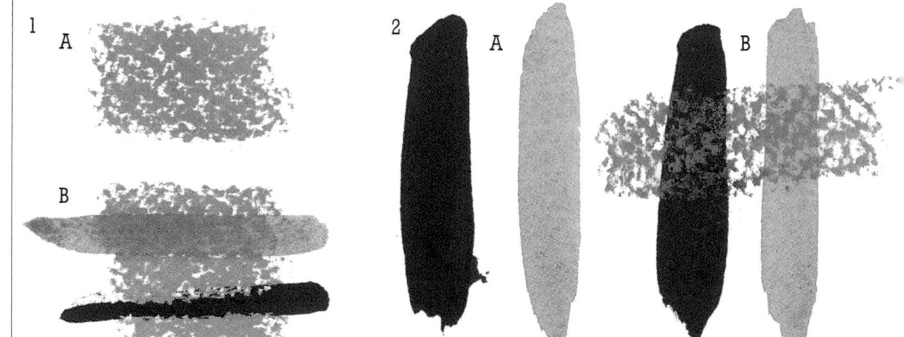

sticks have various properties. When dry, without any kind of alteration, they can be used for drawing in the same way as conventional pastels. They can also be melted by heating. In this way it is possible to create textures that retain their solid appearance when they return to room temperature. Oil sticks, pastels, or crayons can be diluted with turpentine or linseed oil.

By including oil crayons it is possible to achieve very appealing effects.

Oil paint can be mixed with oil crayon or oil pastel because of their spirit-based nature. The oily base creates a matte effect when interacting with oil paint. This effect can introduce very rich contrasts.

Oily and Dry Media: Oil Crayon and Watercolor, the Negative

Oil crayon is a very versatile medium. In addition to the characteristics mentioned in the previous section, it can also be used as a mask. This has the effect of creating a negative of the oily crayon as it causes a water-based media to resist it. Gouache, which is thicker, can cover oil crayon if it is applied heavily enough.

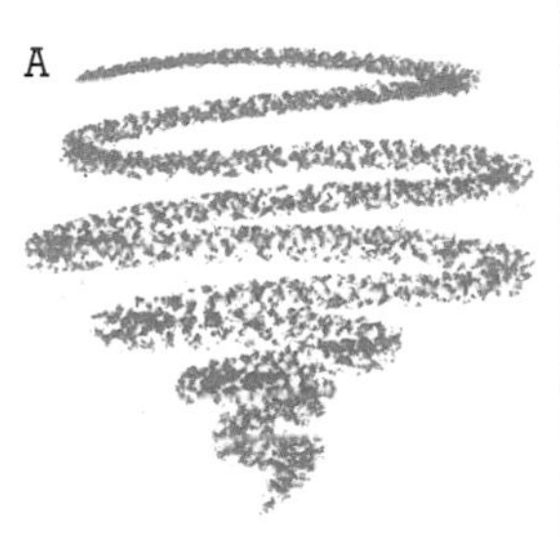

A) Support colored with oil crayon.

B) The negative. A watercolor wash has been applied to sample A.

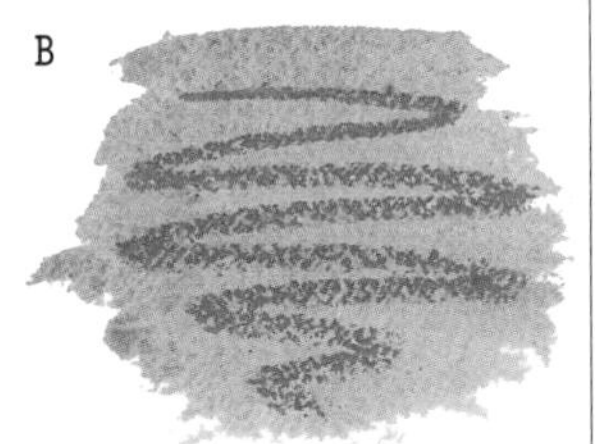

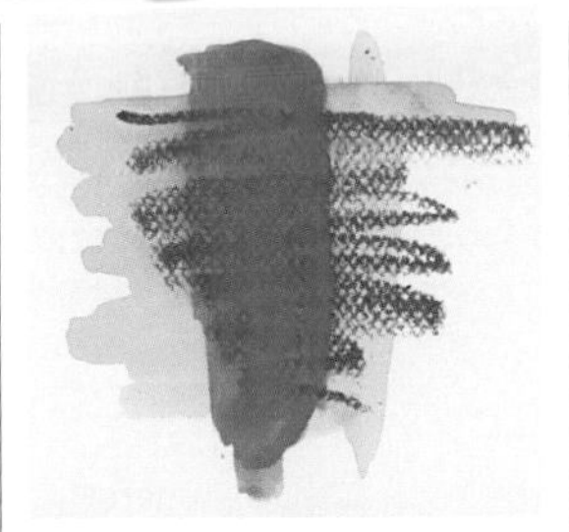

Blues: 1st application of oil crayon, 2nd application of watercolor, 3rd application of gouache.

Here I have taken a sheet of blue, Mi-Teintes Canson paper, colored on its thick-grained side with a dark blue oil crayon. If one applies light cobalt blue acrylic paint, undiluted and straight from the tube, its ability to cover the support and crayon is total.

MORE ON THIS SUBJECT
- Mixed Techniques, 1 **p. 24**
- Compatibility **p. 40**

The acrylic paint adheres well to this base textured with crayon because the support is not completely covered by the oily medium.

Another artist's trick is to use a liquid fixative, which increases adhesive capacity and makes the media compatible.

Here acrylic is used to cover the crayon drawing.

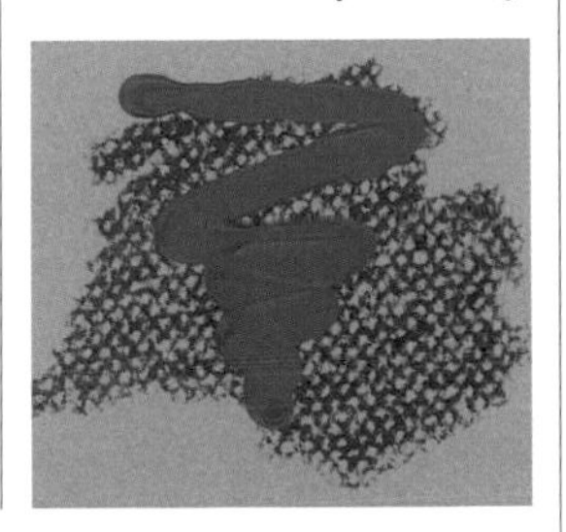

Watercolor and Ink: Two Wet Media

Watercolor and ink are both water-based media. Ink is far thinner and its color potential, though more intense, is more limited in range. Using ink it is possible to draw a sketch, let it dry, then apply washes of watercolor, which the ink will accept without blurring or disappearing. When the watercolor has dried, ink may be applied over it. On the other hand, as the intensity of the ink is far greater than that of the watercolor (even if the ink is very transparent), it will create strong contrasts of tone and gloss.

Used as a conventional mixed technique, applying watercolor with a paintbrush and ink with a pen, one can produce work that is both colored and lined.

When it is diluted, even highly so, ink has the ability to stain with more intensity than watercolor. Furthermore, once it is dry it is indelible. It cannot be worked with water as can watercolor. If washes of colored ink and watercolor are alternated, one can use the characteristics of the two media to produce clean transparencies, whether gloss or matte.

Watercolor and ink.

COMPATIBILITY

Mixed techniques are based on the compatibility of different media. If media naturally reject one another the support itself may permit them to be used together if primed properly. Often this provides new possibilities for creating textures.

Alternating Two Media

Most media can be combined if their particular qualities are carefully taken into account and managed accordingly.

A medium such as pastel is ordinarily used for painting dry. But pastels used with water become a water-soluble medium. As water is the common solvent, pastels and watercolors can be used together.

Again, water is the common solvent of ink and watercolor. The two media can be used in alternate layers. That ink is indelible once dried is a quality to keep in mind as you overlay colors. Remember, the paint can be further worked when dry, the ink cannot. The qualities together offer new possibilities.

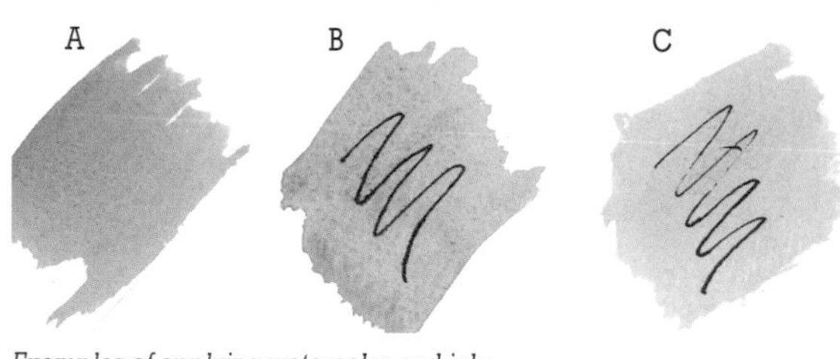

Examples of applying watercolor and ink:
A) Watercolor wash. B) Lines are drawn on the wash using pen and ink. C) Once the sample is dry, water is applied to it. The ink does not smudge, while the watercolor thins and loses intensity.

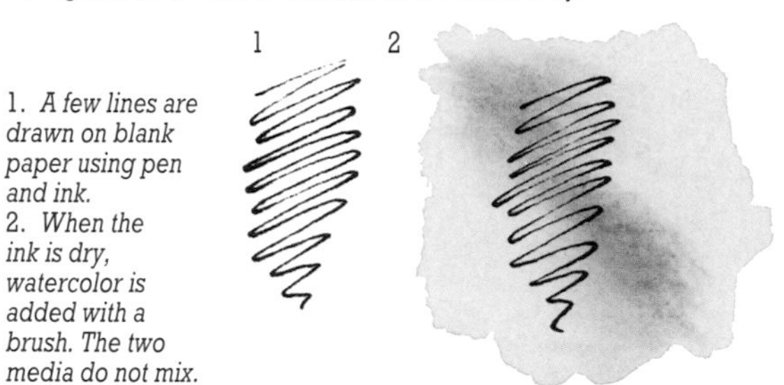

1. *A few lines are drawn on blank paper using pen and ink.*
2. *When the ink is dry, watercolor is added with a brush. The two media do not mix.*

Compatibility Through the Support

If two media repel one another, such as water and oil for example, it is the support to which the paint is applied that allows both media to appear in juxtaposition.

When we talk about using an oily medium such as oil crayon in conjunction with a water-based medium such as watercolor, we are talking about using the capacity of oil to repel, or act as a mask, thus producing a negative. Any parts of the support that are covered with oil will be masked. The watercolor will only color the blank paper.

Therefore the support itself must be capable of accepting each medium.

The matte Schoeller paper that Ferrón uses allows him to submit the finished work to a washing out of its color by holding it under a tap. The strong resistant support of the Schoeller paper allows the achievement of the ethereal finished effect.

Using Figueras paper to paint with watercolors allows the artist to incorporate its canvas-

Miquel Ferrón, Breakers. *A painting with a strong ambience. Felt-tipped pens have been combined with watercolor washes.*

like texture with the transparency of the medium. This paper is also designed for working with oils. Although the operation of alternating the media is laborious, the final chromatic effects are very attractive.

Using the Support for Emphasis

When choosing a thick-grained watercolor paper, or any textured paper that will withstand the application of water, as a support for a mixed technique based on oils and watercolors, the chromatic effect that is achieved is very interesting.

The support permits the juxtaposition of media that repel one another. In this case it is oil and watercolor.

The emphasis of colors through strong contrast.

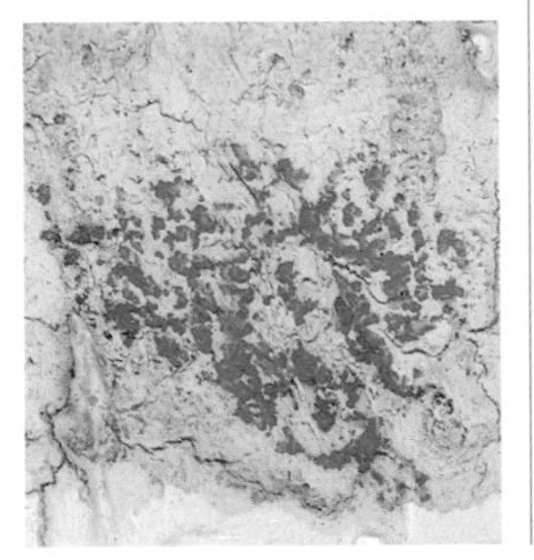

The emphasis is created by the grain of the paper which has the potential to add small, well-dispersed specks of color: oil pigment will cling to the upper part of the grain while a watercolor wash will fill the hollows of the support's texture. This creates an optical mixture.

Using a smooth paper, which increases the transparent, luminescent nature of inks and watercolors, allows the artist to achieve extremely beautiful effects.

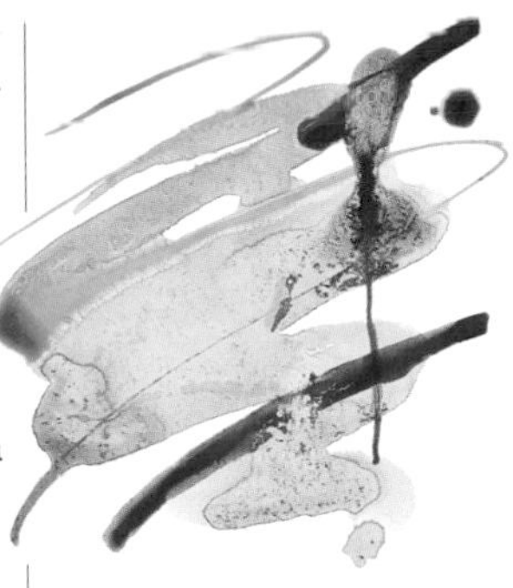

Alternating media and techniques, wet-into-wet, wet-on-dry and wet-into-damp. Inks and watercolors.

MORE ON THIS SUBJECT

- Mixed Techniques, 1 **p. 24**
- Mixed Techniques, 2 **p. 38**

The Versatility of Oil Crayon

Oil crayons are one of the most versatile media available to the artist. As they can be diluted with thinners, they are compatible with oil paint. At the same time, when melted, they can be applied as a liquid, directly onto a canvas or board. In cooling, the medium again hardens, and its texture changes. Either hot or cold, oil crayons can be mixed with oil paints and spirit solvents. In general the crayon gives oil a matte finish. In order to combine these media successfully, it is best to use canvas or board. Then, priming and sealing the support will give one type of effect. An untreated board will offer another, tending to absorb much of the paint's linseed oil, though allowing the crayon to retain its usual look.

Possible ways of handling oil crayon colors:
A) Color applied cold.
B) The same color with a spirit solvent.
C) The same color melted and textured.

WATER: LINE AND COLOR

In any of its states, liquid, solid, or gas, water has some specific characteristics in terms of line and color. As with any subject, it is necessary to observe carefully in order to decipher the nature and direction of its lines, and its coloration before being able to represent it accurately.

A

B

Expanse and limits: A) A river. B) Clouds.

Expanse and Borders

Expanse and borders refers to the horizon line and other elements which act as the water's limits—seashore, riverbanks, etc. The painter will want to learn to depict water ranging from the sea, lakes, reservoirs, rivers, a mountain stream, a simple puddle, or even a snow-covered landscape, as well as clouds and rain, wet streets, and even bottled water.

For any of the states of water in any given landscape, it is necessary to delineate its expanse and its limits. For example, for the sea it would be the lines of the horizon and shore. For a river it would be the banks and the riverbed.

Bottled water has highly defined limits: the walls of the container that holds it.

There are subjects in which the limits of the water are not easily delineated. While in other cases, the delineation in initial sketches is merely suggestive of the space the

Color: It is very important to determine the origin of each reflection. Vincent van Gogh, The Langloise Bridge With Women Washing *(detail).*

The dominant color of this theme is clear.

Choosing a Palette

Whatever the medium that is going to be used, the choice of palette is the first decision. See the examples of some of the most usual ranges of warm colors, cold colors, and broken colors. These are suggestions for oil paints, watercolors, and acrylic.

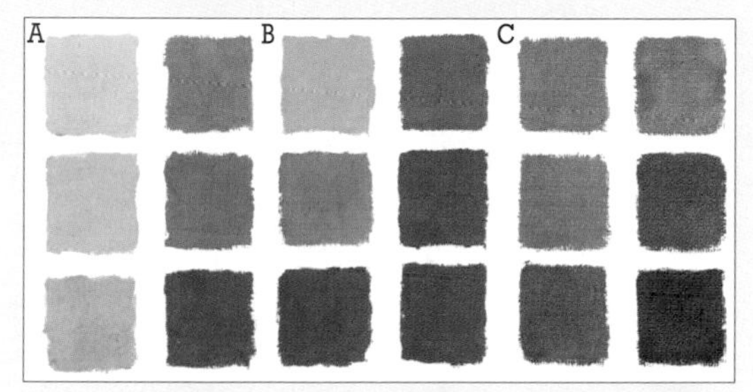

Palette for oil paints:
A) Range of warm colors.
B) Range of cool colors.
C) Range of broken colors.

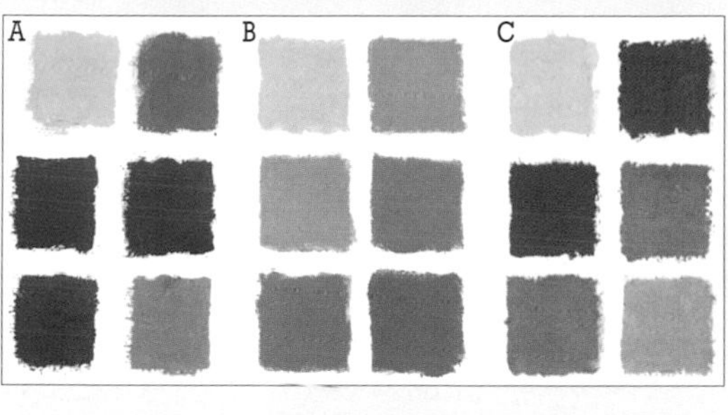

Palette for acrylic paints:
A) Range of warm colors.
B) Range of cool colors.
C) Range of broken colors.

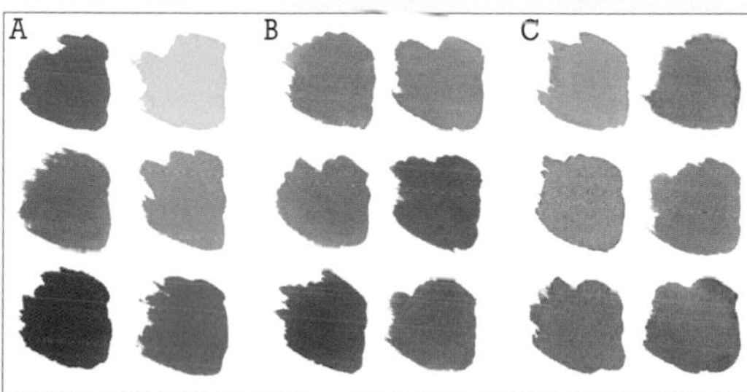

Palette for watercolors:
A) Range of warm colors.
B) Range of cool colors.
C) Range of broken colors.

water occupies. This is particularly true of vaporous clouds or rain or the steam that boiling water gives off. Each of these need special treatment in order to reproduce their effects. For further discussion of this topic, see the subject of textures, on pages 44 and 45.

What Color Is Water?

Besides having to identify the color of water, it is also necessary to discern areas of lights, as well as images reflected in its surface. In fact, if one carefully observes a small expanse of water, it can be readily seen that several elements contribute to its color: ambient light, nearby lights, or lights and darks which are seen through it. The artist looks for an essential characteristic of water in order to lay down a base. To paint it with greater detail, the artist then creates nuance by adding other elements onto this base in subsequent steps.

The Dominant Color

A large surface of water acts as a mirror of the landscape of which it forms a part. Ambient light, possible areas of shine, and the chromatic elements of the surroundings themselves all contribute toward determining a dominant color. Analysis of the dominant color is vital in determining the palette that should be used.

MORE ON THIS SUBJECT

- Water: Texture and Rhythm **p. 44**

WATER: TEXTURE AND RHYTHM

In all its states, water displays forms and expressions which require particular treatment in order to be represented pictorially. The painter can express the movement of water by means of the textures, brushstrokes, and rhythms that are inherent in the application of paint. The construction of texture depends on the handling of the paint and the mix of techniques. The depiction of movement and rhythm is key in the representation of water.

Searching for Rhythm

The movement of waves, the concentric ripples in a reservoir or the flow of a river are very obvious examples of how the subject invites the artist to create rhythms in the texture of the painting. Through both coloring and line it is quite possible to show the movement of water.

For the artist to be able to depict rhythm it is necessary to observe the rhythm and motion of the water. Imitating the water's motion with the brush will give an impression similar to the real image. This creates a synthesis between form and paint. Notice the brushstrokes in the illustrations at the bottom of the page.

Whether an exaggerated representation or one which aims to soften the characteristics of natural movement, this synthesis is always an interpretation created by the painter. But whatever the artistic intention, the image must be sufficiently explicit for the viewer to be able to recognize what is being

A

By observing A) the movement of waves and B) the movement of circular ripples, for example, one can work out its synthesis in paint.

Here are three specific examples of the syntheses of movement in paint.

represented without difficulty. In short, it must communicate the subject's nature.

Rhythm and Texture

All media, techniques, and tools that are used, as well as the support one chooses, offer the artist specific opportunities for creating rhythms. A few examples will illustrate these concepts.

Clouds have to be resolved in subtle, smoothly modeled ways, with the elimination of contrasts that are too harsh and obvious. When working with oil paints, the colors can generally be blended with your fingers. In the case of watercolor, clouds can be painted wet-into-wet, or the outlines of the brushstrokes can be softened once they are dry with a little water applied with a brush or with blotting paper. With both oils and watercolors, in addition to the application of color, the final appearance will depend on how the color is handled to achieve specific effects. Slow transitions in blending color and delicacy of line are key in the representation of clouds.

The handling of rhythm and texture should be modified depending upon how far away the clouds are. Even at infinity, some vaporous masses appear more solid, because they are closer to the viewer. In this case, the relationship between rhythm and texture modified by the depiction of intervening atmosphere will give the painting depth.

The background forms a significant part of a painting and an artist who wishes to master his or her art needs to pay it careful attention.

Clouds in oil.

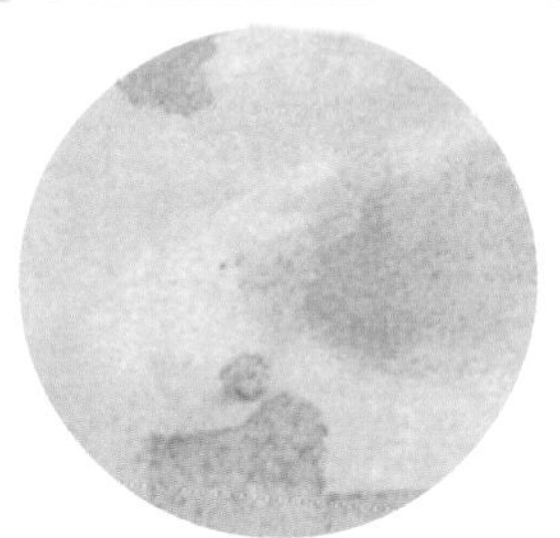
Clouds in watercolor.

MORE ON THIS SUBJECT

• Water: Line and Color **p. 42**

Textures with Acrylic Paint

Acrylic paint offers endless possibilities for creating texture. Here are some examples
a) Acrylic lightly colored with pigment can create hard lines with the help of a paintbrush.
b) On a white background it is possible to texture a very clear wash using purple.
c) This texture is achieved by rubbing a very small amount of paint with an old, dry paintbrush.
d) The effect that can be obtained with different washes is very different depending on whether the base coat is dry or damp. Notice the before and after.
e) An example of wet paint that has been scraped.
f) An example of dry paint that has been scraped.

A

B
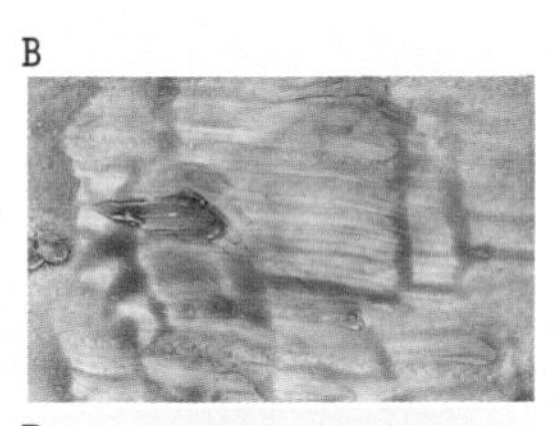

C

D

E

F

Textures with acrylic:
A) An example of texture.
B) A transparent wash.
C) Texture achieved by rubbing.
D) A wash on a dry and damp base.
E) Wet paint that has been scraped.
F) Dry paint that has been scraped.

SUPPORTS

Some supports are exclusively intended for specific media. While other supports might not be, they are compatible and suitable enough. It is also possible to prepare a support by using a primer or by texturing it. In general, though, despite the wide variety, all conventional supports for the majority of techniques can be classified in terms of their rigidity.

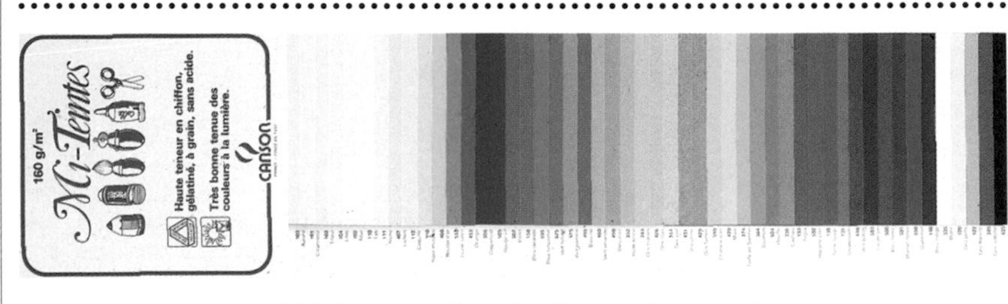

Canson Mi-Teintes paper. A sample of the extensive range of colors.

Paper

Mi-Teintes papers, by Canson, are ideal supports for working with pastels. However these papers will also support a light painting in watercolor or ink applied with pen or brush.

Special watercolor papers, which are usually heavier than those specifically for drawing, can be used with oil crayons, felt-tipped pens and inks.

Geler paper, which is suitable for working with inks, can be used for certain watercolor work, for oil crayons and even for pastels.

The artist must investigate all the possibilities of the paper support and decide which is the most suitable to produce the required textural effects.

Card

Recycled cardboard and corrugated packaging cardboard, all of greater or lesser thickness, are ideal supports for many techniques, whether mixed or not. All that is necessary is to prepare their surfaces. For example, one can prime and texture the surface of cardboard so that the pigment of the media to be used will adhere well.

Other papers: 1) 130g Basik 2) Schoeller 3) Rough watercolor paper 4) 250g matte Geler 5) 250g Montal 6) 240g thick watercolor paper 7) 180g Ingres 8) 160g Canson

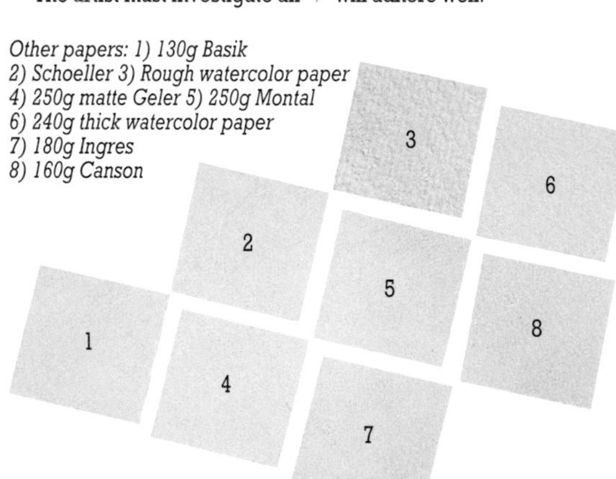

Canvas

Correctly primed canvas is the most usual support for use with oil paints and acrylics. However, thick-grained canvas

Cardboard:
A) Industrial cardboard
B) Corrugated cardboard
C) Covered cardboard

A

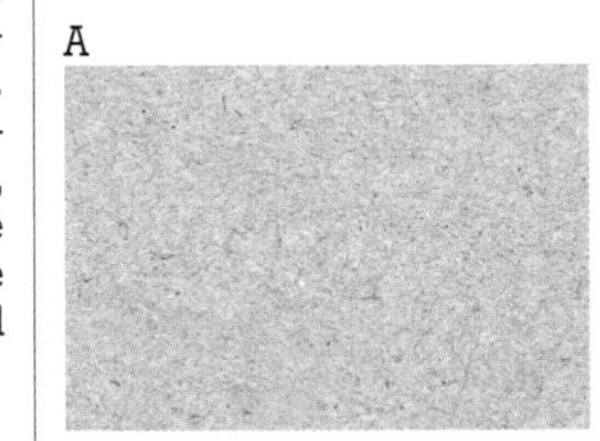

B

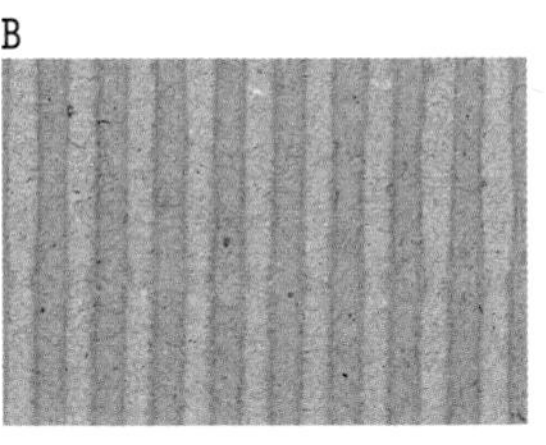

C

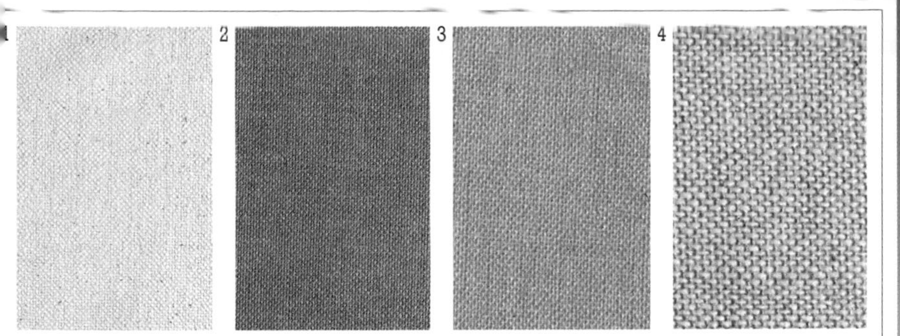

Canvases. There are different types that can be bought by the yard and stretched, much less costly than pre-stretched canvas.

1) Primed cotton
2) Unprimed linen
3) Grained linen
4) The weave and texture of coarse cotton.

has a rough surface that is very good with pastels, oil crayons, charcoal, or pencil.

Thick-grained canvas makes it easier to work with texture, and consequently to create strong contrasts between colors.

Rigid Supports

Wood and cardboard are rigid supports that are available with differing degrees of smoothness or roughness. Oil paint creates a different effect on each of these surfaces.

These rigid supports have to be prepared for the medium which is going to be used and the effect that is sought after. An unprimed board soaks up the oil offering one effect. The result is very different when the same oil color is painted on a sufficiently primed board. Each has its own beauty and strength.

A rigid support does not have to be primed if working with watercolor, gouache, pastels, felt-tipped pens, ink, and so on. But its surface must be sufficiently absorbent to soak up the dampness of water-based media or to retain the pigment of dry or greasy media.

The texture of rigid supports can be incorporated to play an important pictorial role in the finished work.

MORE ON THIS SUBJECT

• Materials and Tools **p. 48**

Laminated wood, chipboard, masonite.

A Textured Support

Acrylic modeling paste is an artist's product that constructs a textured surface to work over, producing exciting effects. It is a material that has many applications, and can be used to create a wide variety of textures ranging from slightly brushed to thick impastos. Once the paste is dry it hardens and becomes very strong, and can be worked on with oils, acrylics, pastels, oil crayons, and charcoal.

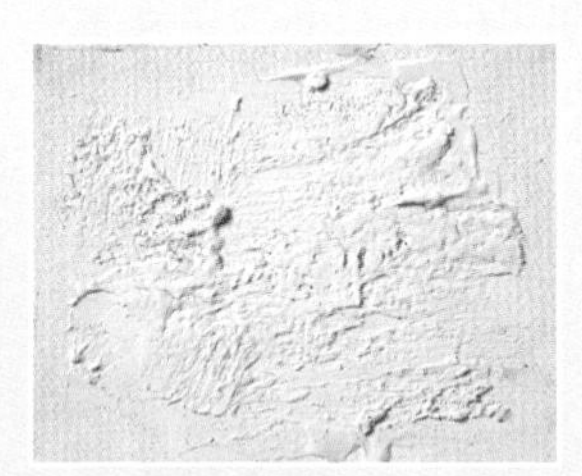

Examples of textured supports.

MATERIALS AND TOOLS

In general, most products related to a specific medium are available in various forms. Some are more suitable for applying a particular technique, others are simply practical to have, and still others hold the appeal of historical tradition. Furthermore, when drawing or painting using a particular technique, it is useful to have on hand materials that will be needed and practical in order to make the painting process easier.

A selection of watercolors.

Specific Material for Each Medium

As an example, let's look at the ingredients of a palette for painting with oils. A universal palette should include, as a minimum, a white, two yellows, two reds, several earth colors including a brown, two or three blues, and one or two greens. In general, most brands offer starter boxes containing a selection of about twelve colors.

In the case of watercolors, one can find small boxes of colors that are very useful for painting outdoors. Larger sets come in both cake form or, similar to oil color sets, in tubes.

A box of oil paints.

For pastels the assortments are varied. The possibilities range from boxes of 12 colors to the hugely extended collection of 522 colors produced by Sennelier.

Basic Drawing Material

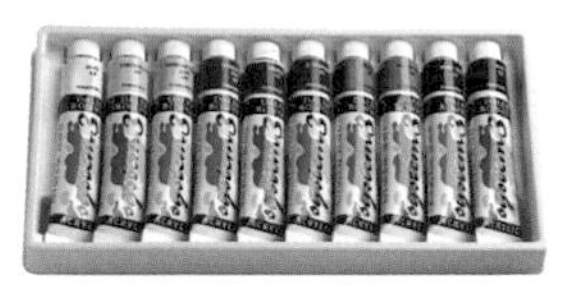

A selection of tubes of acrylic paint.

Wide varieties of sketch materials are readily available. As a material determines the look of a sketch, the painter must decide which one will best begin the piece. The sketch not only sets the tone for further development, but must be incorporated by it. Therefore, its properties have to be considered.

For example, a sketch for painting a seascape on canvas can be executed using a light outline of blues. This color can easily be incorporated into the general tone of the painting. But so can charcoal, as it is delicate and workable.

Watercolor. Paint box (with 15 pans).

Essential Complementary Material

Each medium has its own solvent. For oil paints, linseed oil and turpentine are both necessary. For water-based media (watercolor, gouache, and ink), water is required. For acrylic, in addition to water, gel or another thinner or thickener is needed. For working with alcohol-based felt-tipped pens, alcohol is essential for manipulating the colors while they are wet. To dilute oil crayons, oil solvents (linseed oil and turpentine) are needed.

Paintbrushes and painting knives are necessary for working with oil paint and acrylic. For watercolors, although the paintbrush is traditionally nec-

Basic, common material for sketches.

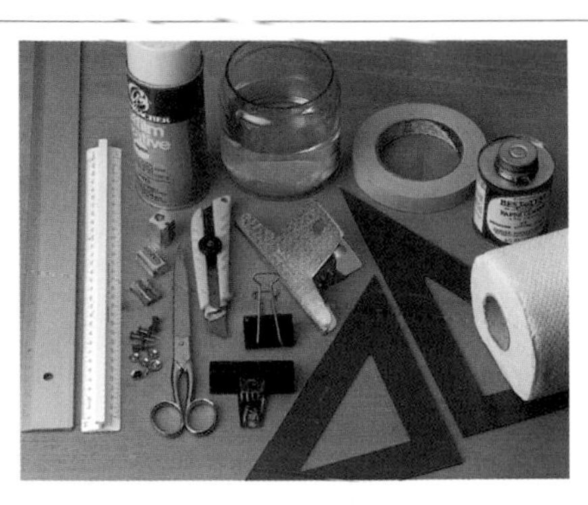

Complementary material for painting with pastels.

All these accessories are useful for working with watercolors. They can also be used for painting with acrylic.

essary, it is not the only tool. Natural sponges or blotting paper and anything else soft and absorbent can be tried.

In summary, a variety of solvents and the tools extend the potential of your medium.

Absorbent papers and blotting papers.

Others

There is a long list of tools which can be used as complementary material. Depending on the imagination and personal resources of the painter, any everyday object can be employed in the service of creating art. Here is a list of some everyday objects with general uses:

A cutting knife can be used for scraping paint (s'graffito), cutting the support, sharpening pencils and crayons, and so on.

Cotton cloths can be used for blending dry media (charcoal and pastels), and for cleaning and drying paintbrushes, both oil and watercolor.

The support. A sheet of paper attached to a board with masking, or paper tape.

MORE ON THIS SUBJECT
• Supports **p. 46**

Blotting paper and paper towels can be used to remove excess paint (watercolor, oils, acrylic).

A palette is essential for painting with oils. But a plate or piece of glass can be substituted, and small containers can be used with watercolor or gouache to make mixtures of colors.

Suggestions for Material for Painting with Watercolors Out-of-Doors

It is useful to have a folder of paper of different sizes and textures for watercolor. The size will depend on the customs and preferences of the painter. For initial sketches you can use a few sheets of notepaper.

As an absorbent paper, paper toweling is very useful, as is a sheet or two of blotting paper.

With regard to the watercolor paint box, one can use a commercially manufactured portable box of pan colors. These may include a paintbrush, a pencil, an eraser, a small recipient for water and a palette for mixing.

Pegs and adhesive tape for securing the paper are vital. The folder can be used as a rigid support, as can a rigid board. The paper must be well secured to avoid unwanted smudges of the wet watercolor.

All of these folders or portfolios would be useful. The size of the folder depends on the sizes of paper being used.

ANALYZING THE PROCESS

Once you have decided on the framing and composition of a given subject, and have identified the different planes and chromatic and tonal areas, it is necessary to think about the sequence of steps to be taken in the painting process. When working with a particular technique or a mixed technique one must follow an order for applying color that is compatible with materials.

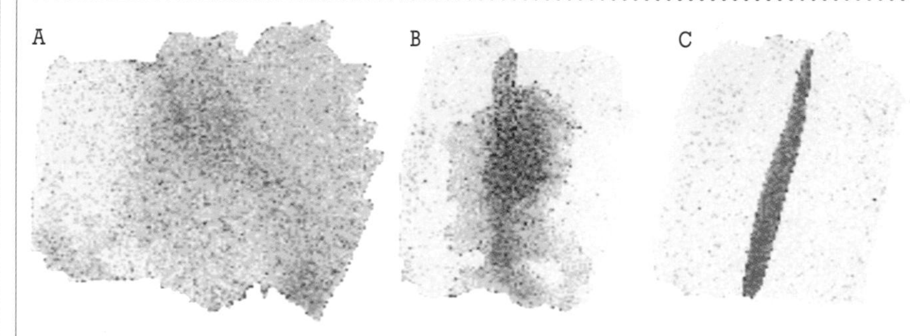

The sequence of procedure using a transparent medium: A) Superimposed coats applied wet-into-wet. B) Superimposed coats applied wet-into-damp. C) A brushstroke applied wet-on-dry.

In this example the difference lies in the time that elapses between the application of each coat. In A) the second coat is applied immediately, without a pause, in B) there is a short time lapse. Meanwhile, in C) the first coat has dried completely before the brushstroke is applied.

Opaque or Transparent Painting

When using the paint transparently, the color that is applied should allow the color of what has already been painted to be seen through it. With transparent mixtures (watercolor, acrylic, oil washes), the color that is applied first needs to be light enough to illuminate the subsequent layers of color. With watercolor, lighter colors are painted first, with darker colors and mixtures being applied and incorporated within the work progressively. Layers of color that are subsequently superimposed, either wet-into-wet, wet-into-damp, or wet-on-dry, will change the color with each layer.

Underpainting

The technique known as underpainting is an exception to this rule of dark on light. An image that is painted initially can be made using a monochromatic tonal range that is then allowed to dry before working over it. The variety of colors (light and dark) that are applied subsequently onto this underpainting, which has been tonally understood will tend to follow the established tonal schema.

When a sketch is made using this technique of underpainting, it can be a very practical tool for speed. An underpainting provides a guide for the overall tonal development of the painting as you add color. The monochromatic notes painted first will

The effect of underpainting with watercolors. In this work the artist does not proceed from light to dark. The sequence is different, and so too is the effect.

offer clues to the appropriate dominant color with which to begin when using the conventional method of dark on light.

Light on Dark and Dark on Light

Paint that is applied opaquely, whether oil, acrylic, gouache, or pastel, can be painted light on dark or dark on light. The sequence in which processes are applied when using opaque paints is freer and less restricted than with a transparent paint.

The versatility of certain media such as oils or acrylics, which can be applied both transparently and opaquely as well as with an impasto, can be complex when it comes to working out the sequence of application of layers of paint.

The Same Goal

Whatever the sequence, it is geared toward an exciting representation of the chosen subject. For this reason, the contrast between colors and tones is key, and strong juxtapositions of light and dark colors is desirable right from the beginning, highly possible when using acrylic, oils, or gouache. On the other hand, with watercolor, acrylic, or water-soluble gouache, the artist can gradually develop successive approximations by first laying down lighter layers, and superimposing darker ones gradually, until the right contrasts have been achieved.

Mixed Techniques

The sequence of laying down strokes and layers when working with a mixed technique depends on the visual and physical characteristics of each of the media used. One must take into account how these media and techniques interact.

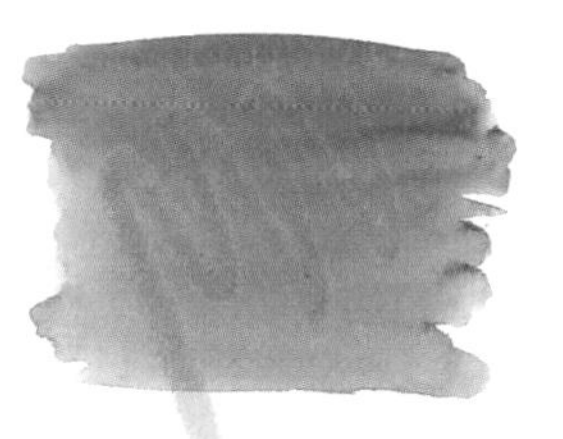

A) On fine-grained watercolor paper, a watercolor wash is applied first. Once it is dry the paint is colored with a subsequent layer of oil crayon.

B) In this example, first the crayon is applied and then the wash. The overall effect in both examples (A and B) is very different.

For example, if the artist uses a mixed technique that includes oil crayon to act as a negative with another medium, it is necessary to apply the crayon first. This holds true when working with watercolor as an area must be masked with crayon before a wash of color is laid over it, causing the effect of resist or color repulsion.

MORE ON THIS SUBJECT

• Watercolor: Resources **p. 32**
• Mixed Techniques, 2 **p. 38**

Alternating Techniques

Artistically enriching contrasts can be created between paint that is applied in highly diluted transparent washes and paint that is applied opaquely. With the first, the texture of the support can be made out through the paint. With the use of opaque paint the support is completely covered. The impression of depth or volume can be developed by using all the techniques available within a given medium. The use of media and sequence will be understood through exploration and experience.

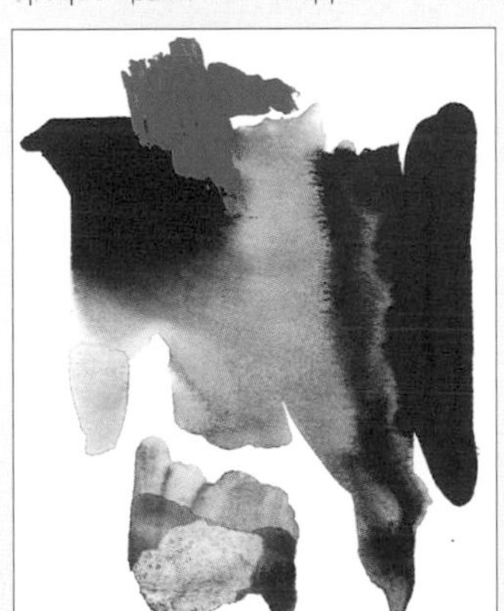

Alternating techniques. In this example the color has been painted on both transparently and opaquely. These effects were achieved with gouache.

THE COLOR OF THE SEA

The color of the sea depends on various factors. The color of the light which illuminates the scene: cloudy, direct sun, late afternoon, moonlight, etc. will be a dominant element. Also objects floating on the surface of the water or which emerge from it contribute to the color. Because of its transparency, one can see seaweed, rocks, and the sandy bottom through its water, and the surface color changes with the water's depth. Because of the sea's great expanse, its color areas will usually cover sizeable sections.

A Mediterranean bay.

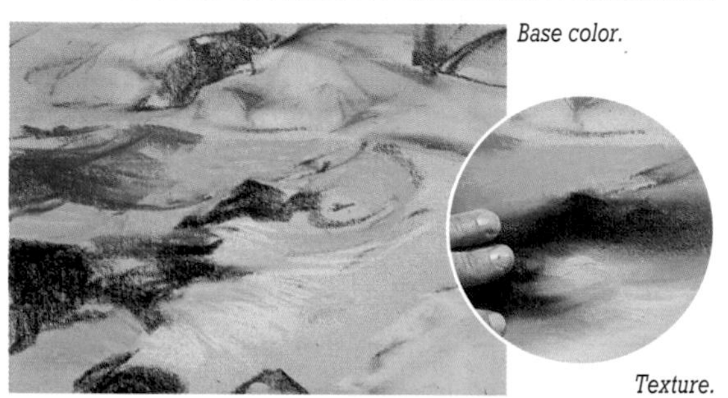

Base color.

Texture.

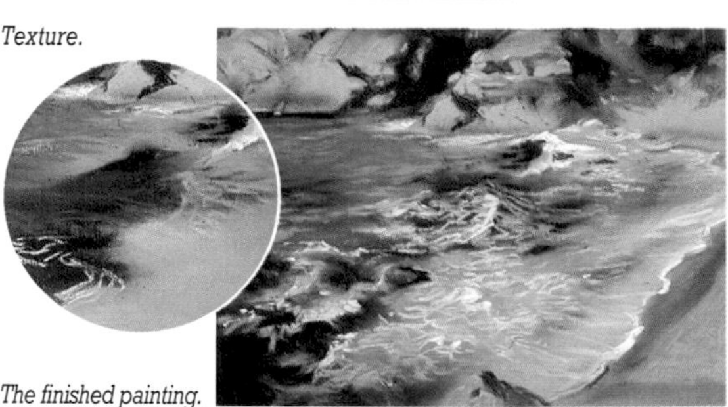

Texture.

The finished painting.

A Mediterranean Bay

Again, the color of the sea largely depends on the characteristics of the light that illuminates it. When the water is crystal clear and therefore transparent, one can make out the seabed through it, with its rocks, seaweed, and sand. But clear or not the body of water takes on the color that results from all these elements.

The small bay in the picture has clear, clean water. Notice the way refracted light shining through the water illuminates the colors of the sand, seaweed, and rocks seen through it.

Painting the sea using pastels can help understand its color.

A sketch is made with reddish brown. Ochres are used for the sand. To represent the sea, emerald greens and blue greens are used, with a few notes of black for the water's darkest parts.

The tones are blended together using one's fingers or a rag. Blending can be used to express the movement of the sea if the strokes follow the water's flow and rhythms.

To this base of color are added touches of white and blue-white. These strokes further define the movement of the water and its foam. The combination of strokes of color and blending can be highly descriptive, hinting that the waves breaking on the sand are flowing back at the same moment.

Rags and fingers are basic tools for the pastel artist. The color is blended and modeled with them except when areas of color and line are left untouched.

A Caribbean Seascape

The elemental color of the sea is based on veridian or thalo green, blue green, and black, with several finishing touches of white and blue-white. But in this Caribbean seascape other colors are used to depict its color.

1) The rocks are colored in ochres and grays. The color of the water begins around the rocks with a very light green and progresses towards a light sky blue. The strokes of color describe the movement of the water. Then the green and blue are gently blended together.

2) Over the base color the texture of foam is drawn, reproducing and synthesizing the

MORE ON THIS SUBJECT

- The Movement of the Sea: Texture **p. 54**
- Foam: Texture and Media **p. 56**
- Reflections in the Sea: Light **p. 58**
- The Sea: Reflected Images **p. 60**
- Reflections in the Sea, in the Foreground **p. 62**

A Caribbean seascape.
1) Base paint.
2) Texture.
3) Finished elements of color and texture.

curves of motion that appear on the water's surface. A few very controlled touches of earthy gray are also applied to suggest contrasts with the sandy bottom.

3) Once everything has been blended to achieve the observed contrasts, the foam can be represented by adding bluish nuances. The impression conveyed in this painting by Ballestar is of the water's motion played against the strong forms of the rocks.

The Color of the Sea in Oils

Each medium has its own techniques. Using pastels, demonstrated in the two examples shown above, Ballestar has created a layer of color that acts as a base for a more nuanced treatment to be applied over it, with details of line and developed contrast. With oils it is helpful to begin with a thin underpainting. This can be developed into a finished painting in one sitting. An underpainting can also be allowed to dry to be developed in successive stages.

1) Photograph of a detail of the sea's surface.
2) For the tone of the base the artist has used a mixture of ultramarine and cerulean blue and white. The darkest area has been given nuance with a little thalo green.
3) To give the sea texture, shadows have been painted onto the base of step 2 to add contrast, with a darker mixture than the base, using veridian green and ochre. The spots of light are represented with touches of white.

THE MOVEMENT OF THE SEA

The painter relies on the textures which can be created by the medium to describe the motion of the sea. We saw in the previous chapter, that by alternating lines and shading, pastels can be blended to give surprisingly realistic effects of the sea in movement. Working with oils or acrylics other types of textures can be achieved. The presence of thicker binders, latex in acrylics, linseed oil in oils, creates the potential for more of an impasto.

Painting a Sea at Low Tide Using Oil on Canvas

This subject with its large clear areas of color is instructive to translate into oil paint. On a drawing that was executed with very few faint lines of charcoal, the colors of the composition are applied using rapid brushstrokes and highly diluted paint. Working in strips from top to bottom the sky is painted first, using a mixture of cerulean, ultramarine, white, and a little raw umber. The mountain is painted with a neutral or broken tone, using the same colors in different proportions. The bluest strip of the sea is painted using cerulean blue and veridian. The rocks are colored with umber. The sea that is closest to the viewer is painted with chromium green, permanent green, ochre, and white. The line of the waves is finished off with cerulean blue and white.

Textures with Oil

Brushstrokes are added to the background color to give nuance to the texture of the sea.

1

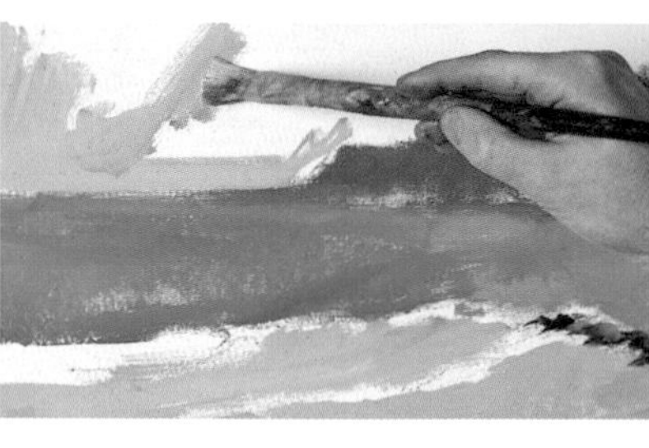

Low tide:

1. *Light coloring.*
2. *Texture.*
3. *Final phase.*

2

3

If one considers the painting as three very different bands, each one is handled in a different way. For the bluest part of the sea, the brushstrokes are horizontal. For the lighter water of the background, the brushstrokes describe the undulating motion of the waves and the impact of light on the foam. For the greenish part of the sea, the waves are painted with diagonal strokes, from top left to bottom right. The dark brushstrokes describe the shadows of the soft waves.

The contrasts created by the details of the different bands give a great impression of depth, as well as making the work expressive and descriptive.

MORE ON THIS SUBJECT

- The Color of the Sea **p. 52**
- Foam: Texture and Media **p. 56**
- Reflections in the Sea: Light **p. 58**
- The Sea: Reflected Images **p. 60**
- Reflections in the Sea, in the Foreground **p. 62**

Seascape with Breakwater, Painted with Acrylic on Paper

This subject was painted using few colors: ochre, white, black, blue, and green. The drawing can be sketched out using a few lines of black paint, establishing the line of the horizon and the breakwater, the edge of the sand meeting the water and the direction of the two waves. These lines are lightened by wiping a damp cloth over them that has been soaked in an acrylic medium and a mixture of blue and black, rubbing in various directions to create a textured base on the paper.

To do the sky, the artist alternated the use of a brush and a painting knife, working in a horizontal direction. The sea is painted next, starting from the horizon, with mixtures of blue and gray (black and white). The color is graded towards green. The paint must also be applied horizontally, except in the description of the horizontal waves. Finally, the brushstrokes and the painting knife impose a texture that represents the soft movement of the waves in the foreground.

Seascape with breakwater:

Watercolor and the Description of Movement

With this watercolor entitled *Lumínica,* Ceferino Oliver has painted a work that beautifully synthesizes light and the movement of water.

Watercolor is a medium that lends itself to painting wet-into-damp or wet-on-dry. The strokes of color of successive layers contrast with the color of the base layer. Edges of the strokes are hard as well as diffused in places. Paint synthesizes the motion of the water.

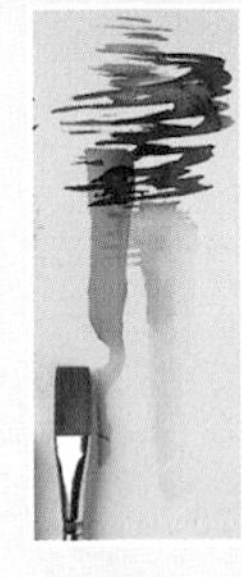

Synthesis of brushstrokes.

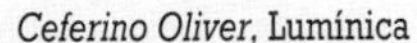

Ceferino Oliver, Lumínica

1. *Adding base color to the initial sketch.*

2. *The sky.*

3. *The sand and the sea.*

4. *Ramón de Jesús,* Seascape with breakwater.

FOAM: TEXTURE AND MEDIA

The power of the sea is one of the most dramatic expressions of nature. Its crash against rocks, strong swells and towering waves are subjects that require more than complex textures. It churns and tosses pulling in two opposing directions, towards the shore and out again. The painter must learn to convey the motion of a tide that happens in a single instant. Capturing the drama of the sea takes observation and experience.

Waves breaking:
1. General coloring using very thin, diluted oil paints.
2. The second session, adding texture to the different planes.

1

2

Foam Using Oils

The depiction of the movement of water against rocks using oils on canvas needs to have the spontaneity of a snapshot. When handling a subject that moves, it is always important to observe it carefully. Study the wave at the moment it breaks against rock. Let your brushstrokes and color masses synthesize what the image expresses. Although a photograph can be of practical help, there is nothing like direct observation.

To paint the foam use mixtures of cerulean blue, thalo blue, and ultramarine blue, heavily lightened with white.

Brushstrokes of the most pure, luminous white with touches of yellow should be used to describe foam. These strokes of white have a strong impact against the base color.

The final touches are added by blending the colors and smoothing the harsh outlines of brushstrokes, thereby expressing the vaporous nature of the foam.

The lightest or most luminous areas should be applied wet-into-wet so the colors may be blended. It is also possible to

Foam and Pastel

Returning to an enlarged detail of the pastel drawing of the Caribbean seascape, one can observe the way the painter uses the strokes of chalk to describe foam on the sea's surface. Each medium has a different capacity to represent a subject. A medium's particular characteristics may, in fact, suggest certain subjects or compositions. Conversely, the nature of a subject may invite treatment by a particular medium.

Enlarged detail of foam on a Caribbean sea, using pastel.

Foamy sea:
1. Foam and watercolor.
2. Initial washes.
3. Final details.

1

2

paint wet-on-dry if you renew the paint around the light areas you wish to blend.

Foam Using Watercolor

The initial sketch is drawn using very faint charcoal lines to situate the forms of the rocks.

To handle the sea, very transparent blue washes are used keeping the brightest areas free of paint. The darkest areas are represented with greenish brushstrokes. Brushstrokes and washes should be applied with the character of the direction and intensity of the sea's movement.

The rocks are painted in dark tones of carmine blue, umber, and some ochre. The beach is painted with a wash of ochre and sienna. The structure of the rocks in the foreground is modeled using wet-on-dry brushstrokes. The shadows define their strong volumes. Their color is intense, with little diluting of the paint.

The sea is handled in two ways. On the one hand white areas are opened up by using water to moisten the dry or damp blues and greens in order to lessen their intensity.

MORE ON THIS SUBJECT

- The Movement of the Sea: Texture **p. 54**

Then, to represent the shadows and reflections of the rocks on the water, umber or Prussian blue brushstrokes are added, wet-into-wet and wet-on-dry.

The contrast between the sea and the rocks is achieved by applying different intensities of color, and using brushstrokes to create texture.

Adding texture: A) On a damp base. B) Creating white areas.

A

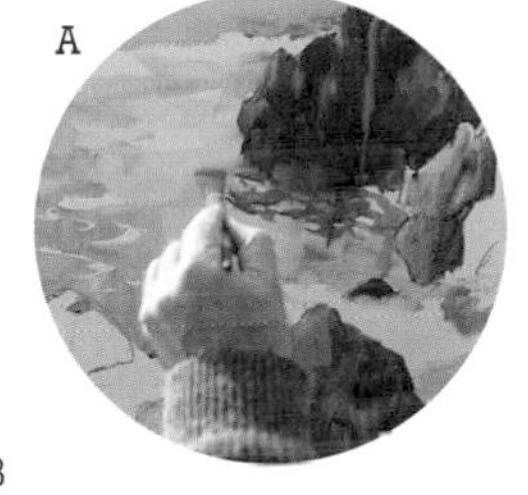

B

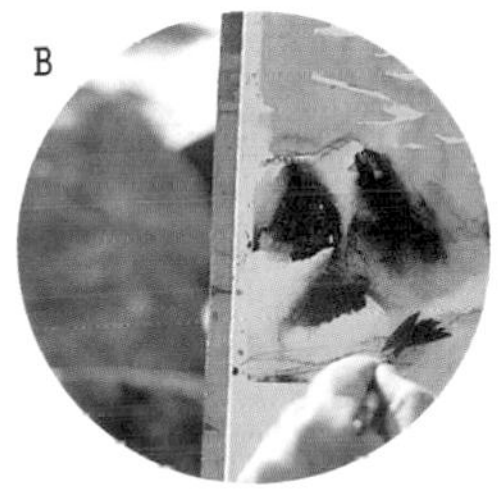

3

REFLECTIONS IN THE SEA: LIGHT

The sea acts like a huge mirror of the sky and atmosphere. A calm sea will produce a very obvious, easily-depicted interplay of light. Even without squinting, which helps to discern the nuances in light, one can perfectly place the light's most intense areas. The specific characteristics of the ambient light will give the subject its dominant coloring.

Sunset.

Dusk.

Observing Phenomena

When the sun appears above the line of the horizon, its reflection on the surface of the sea shines brilliantly. When the water is calm, the direction of the light is clearly vertical and reaches as far as the horizon line.

In the second picture, the sun has already gone down behind the horizon. The sun's rays illuminate the scene with light that produces a general dominance reflected in the surface of the water.

Oil and Reflections

Painting directly onto a canvas that has been prepared with a reddish-gray ground, using a silver white, will establish the shoreline up to the water's edge.

The first patches of color are white, thalo blue, cerulean blue, ultramarine blue, and veridian green. Patches of color are applied in strips, working from top to bottom. The clouds are painted with cerulean blue, a touch of thalo blue, and white. The area that reaches the horizon is painted using veridian and cerulean blue, using the same mixture, adding white for the reflections. The strong shadow of the swell darkened with Prussian blue and umber contrasts with the lighter areas. The water in the foreground is painted using broken white and ultramarine blue. The sand is painted with a mixture of blue and green, chromium green, and ultramarine blue. The base color of the canvas mixes with the dark color of the sand.

A second painting session allows the artist to add texture. The color of the sky is created using white, ultramarine blue, and ochre. The luminosity of the horizon is achieved with a white and a light nuance of yellow. The color of the sea is completed with ultramarine and turquoise, lightened with white. The transparency of the water, shown against the dark shadows of the wave, are indicated with brushstrokes of ultramarine blue. Subtle brushstrokes of white are added and blended in, to depict the texture of the wave. The lines created by the swell are echoed in the foreground.

There is a lesser shine on the sand, which follows the general direction of the reflections that begin at the horizon, continuing to the position of the painter. These are painted using a blend-

Reflections in the sea.

1) Initial coloring.

2) Second session.

A reddish dusk.

1) First, washes are applied wet-into-wet and wet-into-damp.

2) Layers of color are superimposed wet-on-dry.

ing and smearing of ultramarine blue and a lot of white.

The effect that Miquel Ferrón has achieved with this seascape is surprisingly simple. The reflection of light on the sea moves through the whole composition, giving it a strong feeling of luminescence.

Reflections with Watercolor

The reddish light of dusk reflected on a calm sea is a theme that uses a dominant color and will provide us with a complete lesson in watercolor.

A pencil sketch will help guide the artist in the application of color. Luminous colors should be used and dark colors applied on top of lighter ones.

In the background, the horizon is basically painted using wet-into-wet washes on damp paper. Fringes of reddish-blue and yellowish-blue are added. The sea in the background is depicted with a very light grayish-red wash.

The mid-ground is painted using tonal ranges of brown and sienna. It is painted wet-into-wet (the most distant area), wet-into-damp (the intermediate area) and wet-on-dry (the nearest area).

The foreground requires more detail. The reddish reflection of the sea is shown as a gradation of orange as a base, with brushstrokes of blue and umber painted wet-on-dry. The blue of the sea is a wash of blues on a base of red and umber. Next, a blue and umber wash is applied over the area of the sand.

Finally the color of the boats in the foreground is painted wet-on-dry, so that the outlines contrast with the paint beneath, without bleeding or mixing.

The Color of Reflections

This color scheme has been produced using three watercolors on paper: red, yellow, and black. It is a composition with a very light dominant of warm colors. All the mixes were created using these three colors. Mixtures were made to produce oranges, reds, crimson reds, and brownish reds, with the addition of a little black to the crimson.

The color of reflections.

MORE ON THIS SUBJECT

- The Color of the Sea **p. 52**
- Reflections in the Sea: Light **p. 58**
- The Sea: Reflected Images **p. 60**
- Reflections on the Sea, in the Foreground **p. 62**

THE SEA: REFLECTED IMAGES

Within a seascape composition there are two types of chromatic volume that can be of great importance. One is the color of the reflections of ambient light. The other is the effect of the areas of color that represent reflected objects, either floating on the water, emerging from it or close enough to be reflected in it.

In each of these photographs of the port of Barcelona, the specific weight of the compositional balance of reflected images is very different.

Boats in the port.

Oil. Initial coloring.

Coloring the boats.

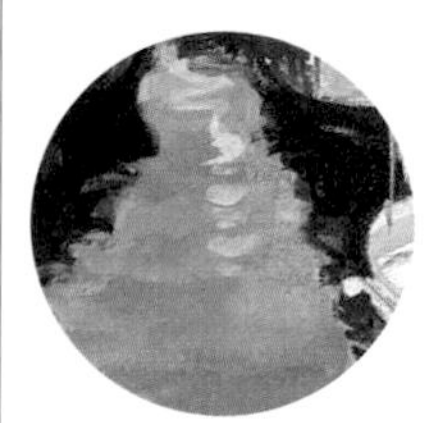

Elaborating and giving texture to the reflections.

The finished painting.

The Chromatic Weight of Reflected Images: Volume and Color

Depending on the artist's point of view, reflected images can acquire real importance in the chromatic balance of the composition. This question is clearly illustrated by the three photographs, each framing the Barcelona marina differently.

The color of the reflected image depends on the object that produces it, on the ambient light, and on the color of the water. A white boat, for example, will give a reflection of pale, broken white with nuances created by all the other elements that appear in the scene.

Boats in Harbor, Using Oils

Take a look at the reflections in the quiet waters of the port, as well as the areas of intense brilliance. The first coloring is made with thinned oil paint. To represent all the water's color areas, greens are used for the darkest, cerulean blue, thalo, and ultramarine for areas that have no reflections, and veridian mixed with ochre for the reflections. To block in the reflection of the boats, blue-green was used (a very dark mixture of ultramarine blue and chromium green). The boats themselves are painted with the same pigments, adding white.

The second layer of paint creates the texture of the sea, the reflected images and the areas of intense shine. In the areas of strong reflection, the texture of the sea is completed with loose horizontal brush-

A snapshot of the port.

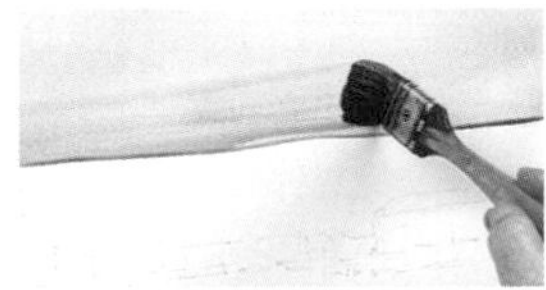

Enlargement of brushstrokes.

Initial washes.

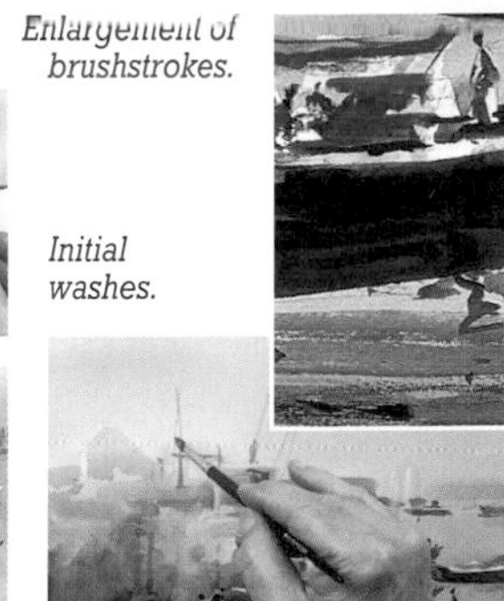

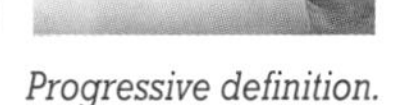

Progressive definition.

Final details.

strokes using a mixture of veridian, chromium green, and a bit of red and ochre to add nuance.

The reflected images of the boats are painted using their corresponding colors tinted with the color of the sea. For the red boat, for example, the reflection has been painted using vermilion, with brushstrokes of ultramarine blue to define it and to mark the reflection from the line of flotation. An outline of the image should appear, either clean or blurred (depending on the water's motion). The shine from the sea should contrast with its first layer of color. The waves are painted using broken brushstrokes of light cerulean blue and white.

Observation

The composition, colors, forms, contrasts and nuance are developed slowly and as a whole, by careful observation of the subject. Each line and color decision is a response to what is seen. The water in the distance, with its texture of smaller brushstrokes, is also less detailed than the mid-ground, where the brushstrokes are larger. The foreground has the most detail and the broadest brushstrokes.

Distortion of the Image

Even a gentle swell can have a distorting effect on the reflections of boats and in this particular case on its occupants. To paint this subject using watercolors, a very light sketch is drawn, and then given a wash with a gradation of grayish-blue for the lower part, warm hues in the upper part. The paper must be left white where the light glints, the points that give off the most light.

While the base washes are still damp, the buildings in the background and the most distant boats are painted, using simple, flat brushstrokes and very subtle contrasts. The color will bleed, expressing the distance very well with highly diluted blues and violets.

Brushstrokes of darker tones of blue and violet are painted in the foreground, wet-on-dry, to describe the flotation line of the boats, the edge of the quay, and the blurred reflections. Two color notes zigzag downwards in ochre and umber to represent the reflections of the two fishermen.

Synthesis with Lyrical Brushstrokes: Egon Schiele

Dramatizing reflections using artistic license is one of the prerogatives of an artist. Reflections are one of the pictorial elements that allow the artist to give free rein to interpretive capabilities. We use Egon Schiele as an example once more (see page 8). The strong, exaggerated lines of the reflections are a powerful element in this work, which was painted on card using oils and colored pencils. The reflections of the masts are stylized and daring.

Egon Schiele, **Sailing boats in water moved by waves** *(detail), oil and pencil on card.*

MORE ON THIS SUBJECT

- The Color of the Sea **p. 52**
- The Movement of the Sea: Texture **p. 54**
- Reflections in the Sea: Light **p. 58**
- Reflections on the Sea, in the Foreground **p. 62**

REFELCTIONS IN THE SEA, IN THE FOREGROUND

A theme of reflections in the foreground allows the artist to elaborate on detail, pulling the composition forward. As far as the composition and chromatic balance are concerned, the reflected volumes have different weights depending on their color and size. Around the reflected image, the water has its particular color as well, depending on ambient light and the colors of the sky.

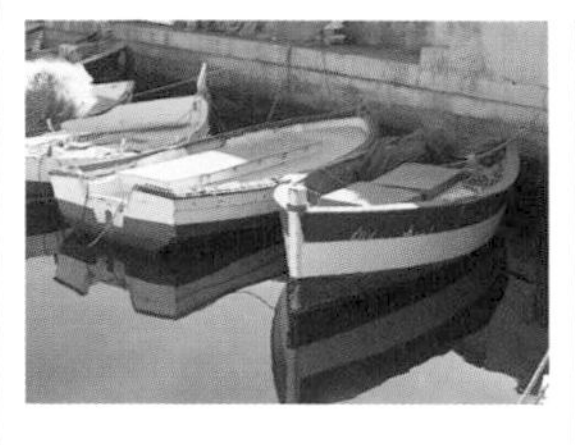

A variety of boats in the harbor.

Ink. A range of colors.

Washes with Colored Inks

In an area of the harbor where the water is calm, a group of boats serve to demonstrate the importance of reflections in a maritime composition. The sun illuminates the scene in such a way that some of the boats are in the shade. There is a clear division of reflections on the surface of the water. They mirror the boats, the ambient light, and the color of the sky. In this diagonal composition, three volumes serve as themes: the boats, the quay, and the water. The latter two delineate the strong line of the boats. The bright blue of the foreground sets off the diagonal of the dark forms.

Range of Colors

The colors used are lemon yellow, carmine (very similar to magenta), two blues, a warm gray, and black. The mixtures of yellow, carmine, and light blue, as the primary colors, allow many other colors to be mixed. The warm gray and the black allow the painter, Miquel Ferrón, to create broken colors with different casts of hue by mixing them with the other colors. The transparency of ink also allows the artist to work with an infinite number of tones of each color, by adding more or less water to the ink.

A simple white china plate can be used for mixing.

The first gray washes on the initial drawing.

Superimposing layers.

The final details in watercolor.

Boats with reflections.

Watercolor. Washes of blue.

Initial Washes on the Sketch

Some of the areas in shadow are colored with highly defined washes made by mixing blue and gray to achieve intensity of hues. The paper is left blank to represent the areas with most light. The reflection of the light on the sea is achieved with a wash that is so diluted that it barely colors the paper (just enough to leave a slight trace).

Blocks of color are added progressively, following the lead of neighboring colors, to create contrasts between the object, in this case the boat, and its reflected image.

It is always necessary to work with highly diluted ink to create transparencies. The ink should remain very watery until the final brushstrokes, which will be darker and more intense.

Superimposed Layers

The paint acquires greater and greater contrast as it incorporates the color of successive layers of superimposed washes. Because of its transparency, the color of the different layers is cumulative. The luminosity of the inks diminishes with each layer, darkening as each one is added.

Inks are insoluble in water once they have dried and so need to be applied carefully, with thought and consideration to color and value. The texture of the different elements of this composition are finally produced by brushstrokes that provide contrast and a greater degree of articulation of detail and volume.

MORE ON THIS SUBJECT

- The Color of the Sea **p. 52**
- The Movement of the Sea: Texture **p. 54**
- Foam: Texture and Media **p. 56**
- Reflections in the Sea: Light **p. 58**
- The Sea: Reflected Images **p. 60**

The Color of the Base

A wash of grays, as in the first exercise, or of blues, as in the second, has a decisive effect on the eventual color that the work will assume as the successive layers of color are added. The medium and chosen palette influence the final result. Whenever one works with transparent media, the underlying coat of paint sets the general tone.

The finished effect.

Final Details

When a painting done with inks has dried, the artist can include definitive details using watercolors or gouache. This work was completed using watercolor. The ropes, for example, are brushstrokes of watercolor that, used in a lighter consistency, add a texture that combines perfectly with the texture of the colored inks.

A Version Using Watercolor

The subject can also be handled using watercolors. Though somewhat more complex in technique, it does not require as many colors. To begin, highly diluted blue washes divide the painting into two areas, light and shade.

The successive applications of washes, using earthy, broken colors, add localized color on top of the initial blue wash.

When all the chromatic volumes have been established and outlined, the final phase of painting can begin. Sabater darkens certain areas in order to create a greater contrast where it is needed. The paints that are used at this stage are very intense and are not very diluted.

LAKES AND PONDS

The water of lakes and ponds moves in very specific ways. There are no waves as there are in the sea. The wind, the rain when it falls, or the passing of a canoe produce very clear ripples on these usually calm waters. These are the textures that the artist must try to reproduce. The color of these waters, being generally still, will depend on the particles suspended in it and on the composition of its bed, on ambient light, and on the sky.

Texturing Calm Water Using Pastels

The base color is laid down on a greenish-gray colored paper. After sketching the outlines in faint black lines, the general direction of the ripples is hinted at using blended blue and green pastel, letting some of the paper's surface show through. Next, more color is added using a greenish-blue, blending it unevenly. By introducing contrasts of blue, the artist begins to represent depth. The colors are further blended to heighten the movement of the water. By adding deep shades of green, blue, and even black, the water is darkened. On this base, all the glittering reflections are added. The touches of light are irregular, loose, and always follow the direction of the lake's gentle ripples. Color has been applied dark onto light and light onto dark.

Reflections on Still Water Using Oils

When there is no breeze, the surface of the water acts as a perfect mirror. For this picture the painter is situated on the lake itself. The reflections of such still water produce an almost identical inverse mirror image of the landscape that creates it. There is a certain amount of distortion in reflections. Here the mountains appear shorter while the trees appear to stretch. The reflected foreground appears much larger, proportionally, since the mountain is seen at a distance. The colors of the inverted images are generally darker shades of the trees and mountains that create them.

The first lines of the composition are painted with thinned oils, using mixtures of ultramarine blue, umber, and white. The horizontal line for the darker vegetation is painted with a mixture of umber and ultramarine blue. The warm, luminescent notes of the trees are

The surface of the water of a lake. The wind creates the particular direction of movement.

Lightening the color base and following the direction of the ripples.

The greenish-blue coloring is textured with irregular strokes of ultramarine blue.

The composition is given greater depth using dark greens.

Where the water sparkles, the color tends toward gray.

painted using orange mixed with white, given nuance and detail by the underlying initial strokes of paint.

Further detail is added to the trees and their reflections with brushstrokes of greens and earth colors. For the movement of the brightest ripples the painter uses the technique of s'graffito on the surface of the lake.

Reflections in the still water of a lake.

The initial coloring painted with thinned oil paint.

To represent the vegetation, individual touches of umber and ultramarine blue are added.

The texture of the ripples and the reflections in the water are finished off using the technique of s'graffito.

The finished effect.

Different Textures

Compare the two enlargements of texture. Each is an example of a different surface, one of a lake and the other of the sea. Both have been produced using the same medium: pastels.

There are different elements that can be used to create texture. Their use is aimed at reproducing the movement of water. Alternating areas of color, line, grading, scraping, and blending will, together with observation, represent the water's surface and motion.

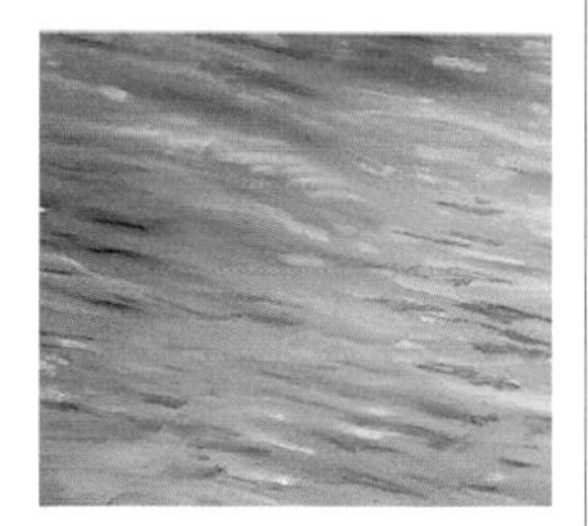

Texture on the surface of a lake, made with pastels. Enlargement of a detail.

Texture of the surface of the sea, made with pastels. Enlargement of a detail.

MORE ON THIS SUBJECT

- The Physical Properties of Water **p. 8**
- The Color of the Sea **p. 52**
- The Movement of the Sea: Texture **p. 54**
- Rivers **p. 66**
- Reflections in Rivers **p. 68**

Gustav Klimt, Lake in Hammer Park

Klimt represents the surface of the water using a beautiful combination of color and texture. The reflected volumes of the vegetation on the river are sensitively expressed. A light breeze acts on the surface of the lake and blurs and distorts the image. There are a myriad of sparkling points of light where the ambient light is reflected, as well as dark, linear nuances in the inverted images of the trees.

Gustav Klimt, Lake in Hammer Park.

RIVERS

The flow of a river, with its bends and turns, its characteristic running water, its vegetation, the artist's point of view, light and silhouettes, possible reflections; these are some of the elements that offer the promise of attractive compositions. Sketches and studies are advisable to explore composition possibilities.

William Russell Flint, A bend in the Seine.

William Turner, Chichester Canal.

The Artist and the Composition

The way the artist composes the scene of landscape and river will largely depend on vision and sensibility.

Here are several examples by painting's great masters. In *A bend in the Seine,* William Russell Flint opts for a composition in which the sky and its light predominate.

In *Chichester Canal,* William Turner captures a magic moment of light and its almost incandescent reflection on the water. Using pastels, Claude Monet offers a quick sketch of *Waterloo Bridge, London.* While on the face of it extremely simple, this pastel drawing is a masterpiece of composition, form, color, and expression. Its effect of luminescence is heightened by the perfectly-defined back lighting, which is emphasized by the color of the ground.

MORE ON THIS SUBJECT

- Reflections in the River **p. 68**

Claude Monet, Waterloo Bridge, London.

Salvador Gonzalez Olmedo. A beautiful composition using verticals, horizontals, and diagonals. The architecture of Notre Dame is painted in juxtaposition with the river and the barge in the foreground.

Compositional Sketches

The painter's point of view of the river, its path, the movement of its waters, will be an important, determinant element in the composition.

Rapid initial sketches are helpful for exploring ways to frame the subject as well as for early problem solving. Those of size, scale, shape, and perspective can be worked out before starting to paint. A sketch using a single color wash will allow the artist to resolve the general form and to block in light and shade using a few sketchy strokes. Sketches will also suggest the next steps necessary to further develop the piece, as will each subsequent step. In the picture by Edward Hopper, there are two curves, one of the river, the other of the road. The repetition makes for a strong, exciting composition.

Edward Hopper, The White River, Foreground.
Adding depth: a more complex compositional scheme.

Different Types of Study

A) A pencil sketch with tonal values locates the important characteristics of the drawing and at the same time allows the artist to identify the subject's darks and lights.
B) A study done with felt-tipped pens and ink. All the volumes are well defined, developed using tonal values.
C) This study was executed in watercolor. Here the aim is to establish coloring. The washes are organized into cool and warm.

Landscape with reflections.

A

Sketch using blue pencil.

Study using felt-tipped pens and ink.

B

Damp techniques. A watercolor study establishing the warm and cool areas.

C

REFLECTIONS IN RIVERS

The distribution of chromatic masses in a landscape with a river running through it depends to a large extent on how much of the composition is reflected in the water. The movement of water is an important element in determining color. Still water can be a perfect mirror. Water in movement, on the other hand, multiplies the reflections and colors.

Autumnal landscape.

Autumnal Landscape

The river in this composition is a secondary element, running diagonally, cutting off the foreground. The river's reflections in the late afternoon light show the inverted images of the trees on the far bank. With the exception of a few points of light that reflect the sky, the whole river is a delicate shade of grayish blue. To be able to achieve harmony of foreground, meadow, and the trees with the mid-ground and the background of grays of different intensities, it is necessary to start with a very delicate palette.

First, wet-into-wet washes are applied using a broad, flat wash brush. With the help of a sponge, the different damp patches are defined. The background, the two banks of the river, and some of the vegetation are established. Next, fine, loose brushstrokes are used wet-on-dry to describe the volumes of tree trunks, the nearer vegetation and the shadows of the trees.

The treatment of the river is left until last, using a very delicate wash of blue, graded to pale as it moves toward background, and used with a little more intensity to represent the reflected images of the trees.

The background is painted wet-into-wet, applying the wash with a broad, flat wash brush.

When the background is dry, the mid-ground can be painted.

More on Reflections

The reflections of trees on the surface of a river after the water level has risen create an interesting interplay of chromatic values. Pastels are one of the most colorful media, creating beautiful chromatic interactions with intense contrasts. This composition is done with a harmonious range of cool colors.

The detail of the vegetation and the water of the river complete the picture.

The images of the trees reflected on the surface of the water are drawn in the foreground. Their detail adds texture.

Pastel is a drawing medium, allowing the artist to make good use of lines to suggest the direction of movement of the water's surface. The lines of reflection are horizontal toward the background, more diagonal in the foreground.

First the colors are blocked in. The contrasts are introduced gradually, adding increasingly dark colors. Blending will create the base color, while the lines that are not blended, which describe points of light, motion, and surface reflection, also texture the water and the trees.

The colors that are used to construct the water's surface, the way they are handled and worked, have the potential to synthesize motion, conveying it by the pastel's final effect. The handling of the subject and medium conveys a moody ambience.

MORE ON THIS SUBJECT

- The Physical Properties of Water **p. 8**
- Rivers **p. 66**

Reflections of trees on the surface of the water.

Creating the base coloring, using few pastel colors.

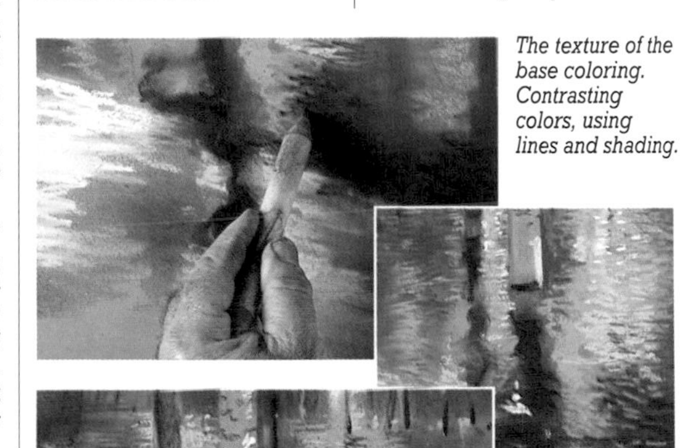

The texture of the base coloring. Contrasting colors, using lines and shading.

The dominant theme of water relies on texture and detail. The artist has captured its essential aspects.

Coloration on the Surface of the Water

Here, reflections and transparent areas have been painted to represent the surface of a river. All the nuances of color and texture give the water the appearance of reality. Notice the chromatic weight of the reflected images against the color of the water's surface. These are distorted, inverse images. The reflections of ambient light and sky provide an exquisite counterpoint to the reflected shapes of land and trees.

Coloration of the surface of a river.

MOUNTAIN STREAMS AND WATERFALLS

Mountain streams, rapids, and waterfalls of any kind all have one characteristic in common, namely falling water and the way it gives movement to the water below as it receives the impact of the fall. The direction of the brushstrokes, lines and movement of the painting knife (when painting in oils) should describe this aspect of the water.

The turbulent water of a mountain stream.

The Texture of Movement as the Water Falls

Observing a continuously falling flow of water, one can identify a series of phases. The fall of water in general describes a vertical or slanting, parabolic curve. This results from the power with which the water reaches the point where it falls and from the particular configuration of the rock or wall it falls past.

The most effective texture for depicting the fall of water consists of lines or brushstrokes which indicate its direction of fall. The color behind and beside it will suggest the surroundings. The direction of lines and brushstrokes, and the handling of the palette knife when using oils, will convey the full force and power of falling water.

The Texture of the Impact of Falling Water

At the point where it falls, the weight of falling water creates an impact that converts it into a white, foaming, vaporous mass. Various techniques can be used to represent this powerful phenomenon. Though lacking lines, masses of vapor assume shapes and forms which can be indicated or modeled with areas of blended pigment. Linear strokes can be used once again to represent the whirlpools and concentric ripples receiving the fall's impact.

The same kinetics of water are what give rise to the direction and form of the brushstrokes, lines, and palette knife movement. A) Direction of splashing water. B) Direction of ripples.

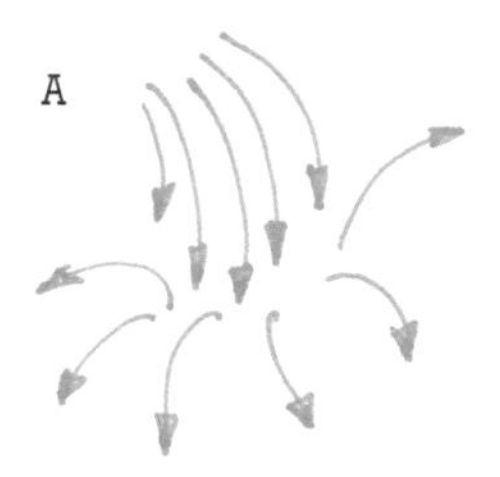

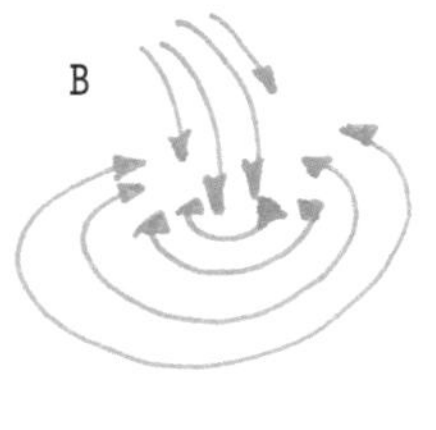

The direction of oil paint blended with the artist's fingers, when the artist wishes to represent specific foaming areas with gentle contrasts. A) Atomized water, in rounded forms. B) Atomized water, in blurred, blended, fragmented forms.

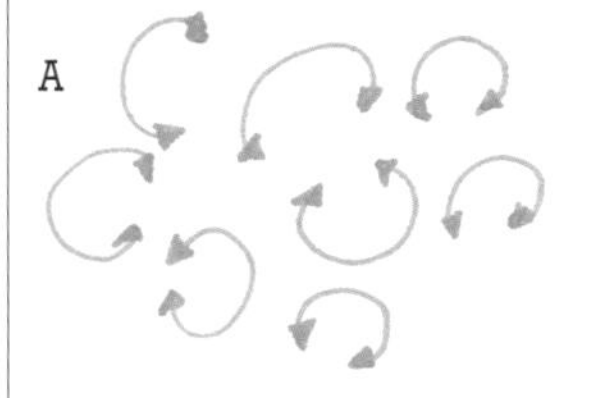

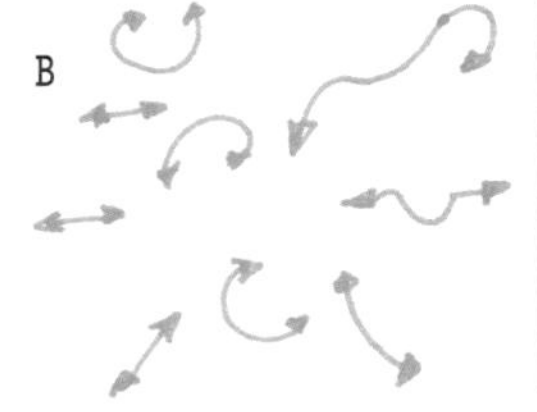

Waterfalls Using Pastels

In this landscape one can clearly see the two phases described above. To express the dynamic of falling movement, the artist has used pastel lines. Then, to depict the vaporous, foaming nature of the water, blending and shading have been used.

The Sketch and Initial Coloring

The most essential forms are blocked out in the sketch. The initial coloring is applied leaving areas of the bluish gray of the paper to represent the medium tones of the rocks. The blues, grays, muddy ochres and whites are used to establish the most important volumes, using blending and shading. A few pure lines begin to describe the movement of the water.

Texture and Movement

Light reflecting from the water falling on the rocks is described with irregular strokes. For the foam which requires the maximum of brightness and

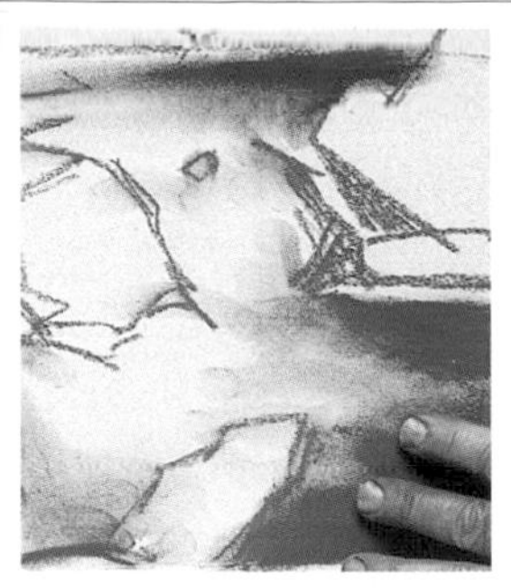

Sketch and initial coloring. The work of blending can be seen here.

Conveying the movement of the water texture.

The water is different in each area. The work is completed by adding points of light and sparkle.

MORE ON THIS SUBJECT

- The Physical Properties of Water **p. 8**
- Water: Texture and Rhythm **p. 44**

sparkle, the base tones are further lightened using a very pale yellow, which, when shaded and blended with the base color, clearly illuminates the white mass of water.

In the next illustration, take note of the strokes of sky blue which describe the movement of the water rushing over the rock to the left.

The final exercise demonstrates how the painter has represented the motion of the water in each space by superimposing finishing touches that add detail to the transparent textures.

Points of light have been added last of all, using individual strokes and touches, to sharpen the illusion of realism.

Understanding Movement

Waterfalls are a phenomenon of gravity combined with the force of flowing water. In the sea, the elements that create movement are winds and tides. These forces act upon the surface of the water. To learn to represent nature in all its manifestations it is necessary to acquire the skill, through practice, of careful observation. Then, using media and the techniques they permit, can nature's subjects be portrayed.

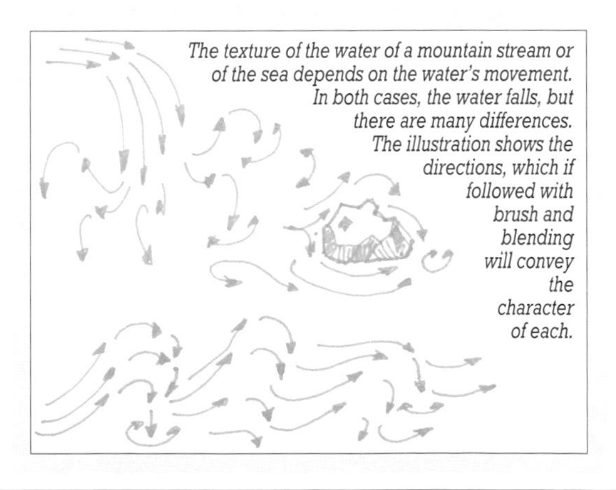

The texture of the water of a mountain stream or of the sea depends on the water's movement. In both cases, the water falls, but there are many differences. The illustration shows the directions, which if followed with brush and blending will convey the character of each.

PUDDLES

The small puddles of water left behind after a rainy day provide a rich element of pictorial interest. The puddles act as small mirrors, in which the sky, ambient light and details of the landscape are reflected. Painting the color of puddles requires careful observation and color mixing.

Drawing Puddles

Puddles are defined by depressions in the landscape. Their surfaces are highly reflective. Furthermore, the transparency of the water creates effects that provide patches of color or tone. A puddle in the foreground which will require detail needs careful observation. The following sections discuss some relevant phenomena.

The Transparency of Water in a Puddle

Puddles are usually quite shallow. When the water is still and any particles in suspension have fallen to the bottom, the color of the surface beneath is visible. The color of a puddle that is not reflecting anything is a shade of the color of the surface beneath.

An example of a composition of puddles. Here they are made by the ruts marked by the passing of vehicles.

The surface of a puddle in the rain is disturbed. In close-up one can see the concentric ripples around each falling raindrop and the small splash it makes. In this case the puddle is unlikely to be clear. It will have a general, overall color.

In the country, puddles are contained by mud or rocks. In the city, they are found in broken pavement and tarmac.

MORE ON THIS SUBJECT
- The Physical Properties of Water p. 8

The reflection in the large puddle in this composition consists of the ambient light. In the small puddle there is a reflected image of the hillside.

Notice the foreground. The color of the puddles is affected by the color at their bottom.

Also, in the foreground, one can see the mud of the track through the transparent water.

The Reflection of Light in a Puddle

When the ambient light is entirely reflected in the surface of a puddle, it creates a wide shining area.

A puddle that has become a bright mirror is an important chromatic element to consider in relation to the balance of the composition. Many artists exaggerate this reflection because it allows them to feature the color white, which can offset areas of color.

Reflected images. One can see a few brushstrokes describing the trees reflected in the quiet waters of these large puddles.

Images Reflected in the Surface of a Puddle

When the water in a puddle is still, one can often see images of nearby objects reflected in it. This can create dramatic and interesting contrasts. Masses of earth or asphalt are juxtaposed with liquid, flowing volumes. Reflected images appear as fragments in their midst. The artist should find points of view and ways of framing the composition that emphasize these unusual effects, taking advantage of details puddles provide when pictured in extreme foregrounds.

Painting outside on a rainy day, or just after it has stopped, will provide ample opportunities to observe and represent the nuances in puddles.

Light reflected in puddles.

Making the Most of the Arrangement of Chromatic Volumes

There are many elements that allow the painter to achieve a strong chromatic balance. In this painting the painter has situated the horizon line in the upper part of the composition, allowing the puddles to assume dramatic importance, structuring the painting with their curves. The composition's big sweep makes for a dynamic landscape.

This watery composition has been painted wet-into-wet and then wet-into-damp. The white of the shine offsets the painting's color.

SNOW AND HOW TO HANDLE IT

Snow is another form of water. It offers an ample repertory of themes for painting, from snowy peaks to landscapes of trees and snow-filled meadows, as well as urban compositions with snow-covered rooftops. The handling of snow's white volumes depends on whether one is using transparent or opaque media.

Snow-capped peaks at dusk.

The sky is painted using wet-into-wet brushstrokes.

The mid-ground areas that are illuminated or in shadow are defined.

Snow-capped Peaks Using Watercolor

The diagonal scheme of this composition is simple to draw. It is primarily an exercise in color. The rocky foreground, which is the darkest plane, delineates the extent of the mid-ground: the snowy mountains. Each of the planes requires a different treatment.

The sky is painted using wet-into-damp washes that leave the edges of the different patches of color soft and blurry.

The snow-covered mountains are painted in stages dark on light. When the sky has dried, base washes are applied with a broad, flat wash brush. First the illuminated faces of the mountains are painted, followed successively by the slopes in shade. The color of the mountains in the background is painted wet-into-wet, with the addition of a few brushstrokes of greater contrast when the paint is dry.

The rocks in the foreground provide the dark notes of the picture, with the most intense color, in order to establish the space and necessary contrasts.

Finally the snowy peaks are shaded by superimposing an earth-colored layer of highly diluted raw umber. In this way the volume is reinforced with a darkening of the areas in the shade.

The strongest contrast is used to represent the dark rocks in the foreground.

MORE ON THIS SUBJECT
- The Physical Properties of Water **p. 8**
- Water: Line and Color **p. 42**
- Water: Texture and Rhythm **p. 44**

The Color of Snow in the Distance

Snow is always described with variations of the color white. However, to well represent it, the use of whites may be quite subtle. Each depiction of snow has its challenges and calls for variations of handling, color sequence, application, and treatment.

Seen from a distance, snow has a general coloring that ambient light colors, both the illuminated expanses and those in shadow.

The task of watercolor or other media used transparently lies in applying initial washes that clearly convey the definitive color of areas of light. The challenge of oil paint lies in learning to keep the color white intact, so that it does not degenerate into unwanted grays. Brushstrokes used for the whitest, most luminous areas should not be applied at the beginning, but rather saved for the end.

A snow-filled landscape.

The initial coloring is completed using cool colors.

Charcoal sketch and initial coloring using warm colors.

The contrast of the dark foreground delineates and marks out the mid-ground and the distance.

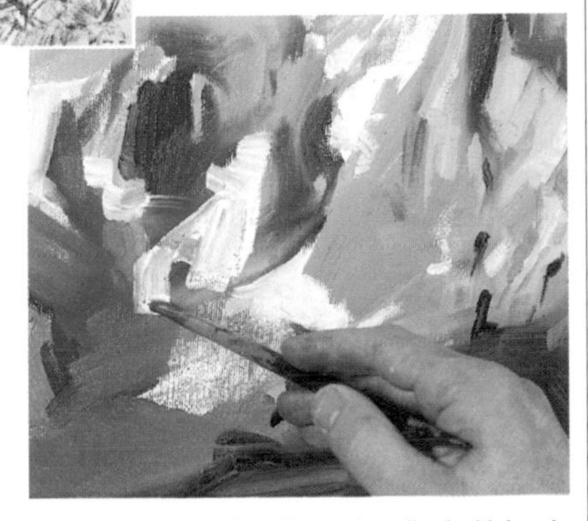

The snow-covered peaks are described with brushstrokes of white, or whites mixed with other hues.

A Snow-capped Mountain in Oils

With a medium like oil, used opaquely, this snow-filled landscape offers a lot of possibilities. The very dark foreground meets and defines the mid-ground. The sky, with its clouds obscuring the peak of the mountain, provide a counterpoint by softening the sharp line formed by the crest of the mountain range.

After producing a rapid sketch in charcoal, the first coloring and base tones of the mountain can be applied. These are mixes of ochre with a little umber. To depict the mountainsides in shadow, an ultramarine blue is added, mixed with white, and with a little carmine, so that the shadows appear more purplish.

The sky is painted with ultramarine blue and white in its upper part, changing to cerulean blue with white in the lower part, where it touches the peaks. The volumes of the clouds are painted with white, blended, using the same brush, with the blue of the sky which is still oily and workable. The paint of the mountaintop is also blended wet with the paint of the cloud, so that volumes seem to emerge from atmosphere.

There are areas of the mountain, namely the snow fields, where the canvas is still unpainted. These require a careful treatment.

Brushstrokes of white paint applied to the blank canvas, or onto underlying paint, should be confident, descriptive of volume. Even when scrubbing away color, lightened areas need to describe form. A little ochre or blue is added to the white paint to give shadow and nuance to the whiteness of the snow where appropriate.

Blended brushstrokes show the vaporous nature of the clouds clinging to the mountain peaks.

Gustave Caillebotte, Vue des Toits (Effet du Neige). *A detail of a scene of snow-covered rooftops.*

SNOW: COLOR

Snow-covered mountains can form part of the distant backdrop of a landscape. To represent them, each medium has its own techniques. The handling of paint changes depending on the nearness or distance of snow. In this chapter we will also look at creating atmospheric effects using watercolors.

The Color of Snow

Seen from close-up, snow has specific chromatic characteristics. Untouched snow, snow that has been walked on, compacted snow, or snow that is dirty from mud all have different colors and textures.

It is important to learn to paint with whites. To represent snow in all its variations it is necessary to understand how to mix them. Broken or mixed whites are grayish whites, bluish whites, and earthy whites.

Depending on the medium with which the snow is painted, the sequence of paint application and handling are different. With an opaque medium, the varying shades can be applied directly. But with a medium that is applied transparently it is necessary to leave areas of clear white, and to apply dark color onto lighter colors. It is its coloration that conveys the texture of the snow.

Landscape with snow.

The first wash preserves the white of the snow.

Observing the Nature of Snow

When one observes snow at close range, one can discern a great number of hues. In the photograph on the opposite page, the curving track introduces a diagonal, the tree in the foreground is vertical. Both provide a contrast to the line of trees on the horizon in the background. In this painting, by Ballestar, the snow creates a triangle, modified by the colors and shapes of landscape and shadow, and balanced by the sky.

The artist paints the color of the sky, not quite, but almost white by applying a highly diluted gray wash that leaves a faint hint of color on the paper.

Later brushstrokes continue to support the white of the paper.

The painting progresses, leaving the white areas as they are.

This near-finished study shows the white areas of paper in contrast to the strength of the rest of the painting. These are whites that describe the condition of the snow.

A track in a snow-filled wood.

Description of planes using a wet-into-wet technique in the distance and wet-into-damp or wet-on-dry in the foreground.

The color of the sky.

The vegetation and texture of the snow are painted at the end. Notice that there are still areas of perfect whiteness on the paper or support.

The luminosity of the sky painted with watercolors sets off the snowy landscape. The cold, icy surrounding atmosphere is represented by the cool, highlighted, gray wash.

The work begins by reserving the areas that will remain white. The nuances of color have been created using very dilute brushstrokes of gray of varying intensities, superimposed on one another. Small touches of umber and sienna are applied by blending color using wet-into-wet, as well as painting wet-on-dry, for touches of emphasis in the large volume of shadowed snow.

MORE ON THIS SUBJECT

• The Theory of Color **p. 18**

Snow and Distance

What the human eye can perceive depends on how far away the object being observed is. Foreground images will contain more color change and nuance, as well as indications of textural detail. Mid-ground becomes less specific.

Snowy peaks seen from a distance become simpler areas of color, impressions of wide, snowy expanses. Snow viewed from far away appears as clearly delineated areas of light and shade.

Blue landscape. *One can see the very clear contrasts between areas of light and shade in this snow-filled landscape.*

ICE

Because of its polymorphic or changeable nature, water appears in solid form under different guises. Water in the form of frost is crystalized and creates myriad points of light. Ice, unlike snow, is transparent. In a composition featuring ice, reflections and refractions can occur.

The Shape of Ice

A few simple ice cubes can act as an example. In this example, the outline of the ice is very simple. Ice cubes take the shape of the mold in which they are made.

In nature, water at low temperatures appears as frost, hail, and icicles. It occurs as frozen ponds, lakes or reservoirs, icebergs at sea, frozen seas at the poles, and glaciers.

As frost, the dimensions of the ice crystals are so small that from a pictorial point of view its presence is translated into points of light.

The shapes and forms in which water appears in its frozen state are varied. Hail, which falls like rain, forms small spheres which range in size from very small to considerably large. Icicles have conical, layered forms. Large surfaces of frozen water are also layered, formed unevenly, and translucent in appearance. Icebergs and frozen seas form veritable mountains of ice with dramatic shadows and multiple hues.

William Turner, The Frozen Sea.

Turner and Glaciers

The Frozen Sea by William Turner is a beautiful watercolor, its composition constructed by large, dramatic forms. Jagged ice mountains tower from a landscape of snow. Notice the way the artist has represented the characters of ice and snow, how clearly different they appear from one another. The broken line of the blocks of ice easily distinguishes them from the shadow-filled landscape, where the outlines are rounded.

Transparency Through Ice

Ice is transparent. Colors and forms behind or within it can be seen in distorted images. This creates some interesting effects and nuances.

These photographs of ice illustrate its transparencies, reflections, and distortions.

Reflections

The surface of the frozen water of a pond, puddle, lake, or reservoir will reflect ambient light.

When ice is smooth and finely formed, even images of landscape are reflected by it. The quality of the image can be almost as precise as the surface of a mirror.

As ice, water can take many different forms. The form of ice cubes will depend on the shape of the mold.

MORE ON THIS SUBJECT

- Water: Line and Color **p. 42**
- Water: Texture and Rhythm **p. 44**

Photograph of ice, with colorful objects visible through it.

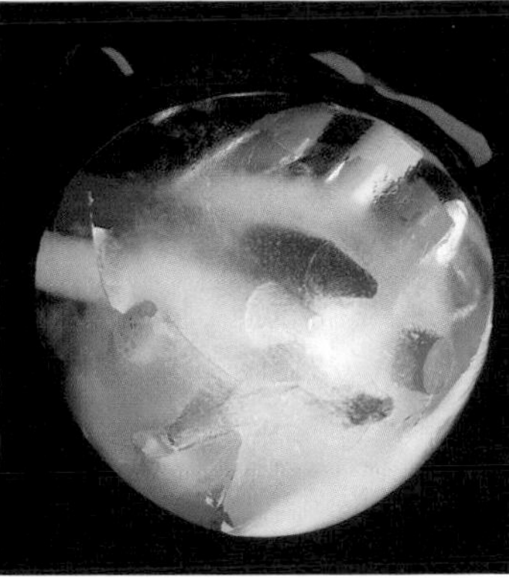

Photograph of an ice cube with an object frozen inside it.

Photograph of objects and their images reflected in a block of ice.

Learning to Analyze

Ice, when viewed in the foreground of a landscape, is full of nuance. The colors used to represent ice should be mixed by analyzing all the elements present. First, the ice will have the coloration of the water which formed it, (bluish, aqua, muddy, etc.). It will also take on the colors of the elements surrounding it, as well as the objects frozen into it.

Brueghel, Hunters in the Snow. *Although they are in the mid-ground, the images of figures reflected in the ice can be seen faintly.*

Snow and Ice

In landscapes in which snow and ice are represented, the contrast of texture and color marks the difference between them. Medium and technique must be worked to support the transparency and solidity of ice and the opacity and softness of snow.

Brueghel, Landscape with Skaters and Birdtrap.

CLOUDS PAINTED WITH WATERCOLOR

Vapor is another state which water assumes. The substance of clouds, vapor often plays a dominant role in landscape composition. Clouds frequently extend as far as the eye can see, providing the content of background to land and sea and their features. Though clouds have definite forms their edges demand undefined treatment, almost without contrast to adjacent colors. Every medium has its techniques which will express the nature of clouds.

Vapor Using Watercolor

For the most distant clouds it is a good idea to apply a very diluted wash, wet-on-wet, having previously dampened the paper in the area of sky. Next, remove color and moisture, lightening the forms to express their mass, using a semi-dry sponge. When the paint is dry the color will appear free of hard edges.

A B

Examples of skies painted with watercolors.
A) The cloud's edges are soft, undefined (wet-into-wet washes).
B) More definition in the outline of the clouds (painted with less moisture).

Creating White Spaces

White spaces can be created in many ways while paint is still wet. With the help of blotting paper or paper toweling, the edges of the dry area can be preserved. A clean, dry paintbrush permits the removal of some of the dampness and the excess color before areas have dried.

Creating white spaces with watercolors that have already dried is also always possible. With a clean paintbrush saturated with water the area where the artist wishes to lighten the color can be moistened. After allowing the moisture to work into the area, excess color can be removed using blotting paper or a clean, dry paintbrush.

By alternating all these possibilities one can achieve different levels of definition at the edges of areas of color.

Notice the variations of definition possible working in this way. The final photograph on this page shows how it is possible to represent a cloudy sky, and lighten it by thinning the color using simple brushstrokes to re-moisten the paint.

Always Wet-into-Wet

Because clouds appear at a great distance, the outlines should not be defined clearly. The following procedure will

A group of clouds made by opening up white areas on dry watercolor.
A) A gradation of blues that has dried.
B) Recovering the whiteness of the paper. The painting is moistened with water where the coloring needs to be thinned.
C) Brushstrokes of sienna, umber or Payne's gray are painted on the resulting white areas, to complete the vaporous nature of the clouds.

A

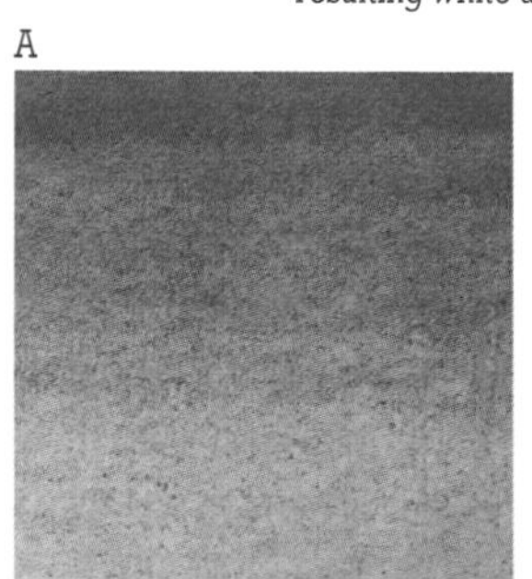

B

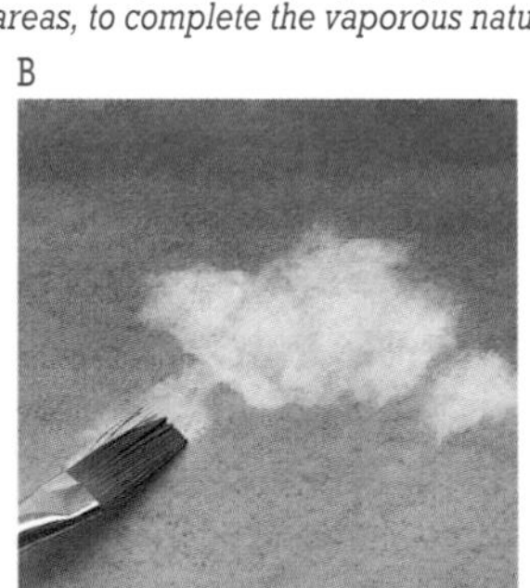

C

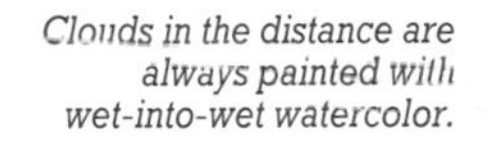

Clouds in the distance are always painted with wet-into-wet watercolor.

Transparent watercolor lends itself to the depiction of luminous skies.

help you achieve the illusion of distance. Although it is possible to thin the color of clouds once the paint is dry, any new application should be made wet-into-wet, to keep the edges feathery.

Using and Maximizing the Flow of Color

When working with watercolor, the support can be tilted to utilize the pull of gravity. By tipping the work, first one way then another, the flow of paint can be guided. Interesting shapes, curves, and linear effects can be produced by tilting, adding freshness and immediacy to the painting.

The luminosity or sense of the presence of light and clarity of color in the finished work depends on the handling of its layers of paint.

In this landscape the color is clear and luminous. The composition is full of strong contrasts and the illusion of light.

The expressive possibilities of allowing watercolor to flow.

Absorbing Excess Color and Moisture

Absorbing excess color and moisture with blotting paper has a different effect than when removed using a clean, dry paintbrush. Blotting can achieve a free, loose effect, while the brush must be handled with control, lifting color carefully to avoid leaving traces of unwanted paint.

Lifting color to achieve form using blotting paper can be difficult, particularly because contrasts created by blotting paper are much harder. Pigment often bleeds deeply into the grain of the paper.

Removing color and moisture with a paintbrush permits more control, though its handling takes practice and experience. The use of a sponge, blotting paper or paper toweling can, however, produce highly textured, cloud-like effects.

MORE ON THIS SUBJECT

- Water: Line and Color **p. 42**
- Water: Texture and Rhythm **p. 44**

CLOUDS PAINTED WITH OPAQUE MEDIA

Whatever the medium chosen to represent clouds, it is necessary to express their vaporous consistency. Pastel, with its capacity for blending and shading, is an ideal medium for painting clouds. Oil paint used either opaquely, even with an impasto, or used transparently, is another of the great pictorial media with which the artist can represent cloudy volumes.

Clouds Painted with Pastels

Techniques for representing clouds using pastels utilize the medium's soft, smudgy look. The edges of clouds, even when clearly defined, need to be light and blended. Clouds are huge, billowing volumes of water vapor. Pastels naturally create misty, diaphanous layers, uniquely suited to creating the illusion of cloud forms.

The Texture of Clouds

In order to express the texture of cloud vapor pastel should be handled with gentle color gradations and soft, thin, blended edges. Clouds are the most distant forms and volumes the human eye can read, lying almost at the limit of our vision, as far as the eye can see. They must always be softly depicted, worked as general masses that may have substantial volume but lack solid consistency.

The clear skies that surround clouds also require soft color gradations, of violet blues and green blues, reds and even pale greens for sunrise and sunset.

Cloudy skies painted with pastels. The dominant colors are different. The potential for chromatic richness is infinite.

Skies with large groups of clouds.

A reddish sky.

A nocturnal, stormy sky.

A Changing Scene

A sky with clouds is constantly changing. The artist interprets what he or she sees. The task of the painter is to shape and mold the composition so as to express the major characteristics of this changing view. The painter conveys the impression of having captured a brief moment, representing the fleeting shapes of clouds, their colors, intonations, areas of shade and the light that shines through and illuminates them.

Constant Elements

To express the volume of a vaporous mass it is necessary to express the intensity and effects of light that appears behind it or that shines through it.

This is partly accomplished by laying in a background color, its essential tones representing the sky. Cloud masses can then be applied to this base. Different degrees of contrast will help to give the impression that some clouds are further away than others.

The sky can be organized into foreground, mid-ground and background, three general spaces, much the same as a landscape. Laying in the color of the background provides a base on which the rest of the painting is worked. The observed shapes and colors of clouds painted with layers of transparent color, as well as opaque areas, will gradually create the impression of vaporous volumes.

Blending Oils with Brush or Hands

When painting clouds with oils, their vaporous forms can be captured by blending the paint, or working it with the fingers.

Colors brushed in next to each other can be gone over either with a clean, dry paintbrush or the fingertips, to smooth transitions and eliminate sudden leaps in color or tone.

By blending with the fingers, one can apply small amounts of color to an area of paint that has dried. Although this does not give as vaporous an effect as blending wet paint, it allows the artist to alter color and introduce nuance without reworking the whole surface.

An example of oil blended to give the texture of a cloud.

An example of working paint with the fingers. A different texture is obtained.

MORE ON THIS SUBJECT

• Analyzing the Process **p. 50**

Palette Knives and Oil Impasto

A thick surface of paint can be applied and textured with a palette knife. Such an impasto can be highly expressive of clouds with all their motion and rhythm.

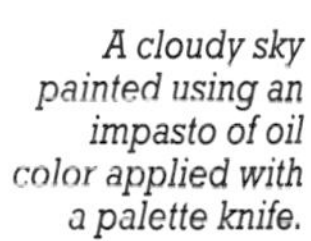

A cloudy sky painted using an impasto of oil color applied with a palette knife.

Vincent van Gogh, Cypresses. *The clouds are textured with a thick impasto.*

The Color of Clouds, an Unlimited Palette

The color of the sky and clouds depends on the color of the light. Furthermore this shifts and changes depending on the time of day, the season of the year and the weather conditions.

A cloud-filled sky can take on a black or intense, midnight blue appearance; it can look yellow, gray, or pale blue at dawn; or orange and purplish red at sunset. Clouds can be white, gray, or brown. The color of light that lights a cloud-filled sky calls for a full palette of colors.

Clouds overlap and hide one another from view. They throw shadows. These cloud shadows are cast on hillsides, mountains, and buildings. Their volumes and forms are some of the most beautiful forms in nature. Their contrasts and colors can add majesty to a composition.

Notice the shadows that the clouds project on the surface of the earth.

FOG AND MIST

Fog and mist are also forms of water. Whereas clouds have clear outlines, fog and mist produce soft blanket-like forms that float and veil the landscape. Natural volumes are softened and grayed when seen through the pale, moist atmosphere of mist or fog.

Mist Painted with Pastels

Pastels are the medium par excellence for producing soft, atmospheric gradations and blends. The ethereal warmth of the effects produced by pastel allows the artist to express the gentle, blurred forms of a foggy landscape with great realism.

In his work *Marine-Soleil,* Degas, a great master of pastels, expresses with soft, luminescent beauty, both warm and intense, a misty landscape at dawn.

Mist Painted with Watercolor

Watercolor is also well suited to the representation of landscapes in fog. Highly diluted washes, applied in delicate overlays can produce the soft forms and atmospheric illusions created by fog.

To begin a painting of a landscape seen through mist or fog, lay in a highly diluted wash of greenish-gray. Blue brushstrokes can then be painted onto this wash to describe the mountain ranges. The atmospheric effect is achieved by applying the colors in very thinned washes, painting wet-

Watercolor Landscape of Snow and Mist

To obtain a general tone in a misty landscape, a wash is applied to represent the sky and the ambient light. Here the sponge is a useful tool. It can be used to cover the entire surface of the paper with just a few sweeps. With a sponge the artist can also allow more moisture to remain in a particular area as well as remove color and moisture where needed.

This landscape owes its general grayish intonation to washes composed of very few colors: sienna, umber, green, and Payne's gray.

The atmosphere is achieved using watercolor washes. Here the washes of more than one color allow the artist to paint a work of very subtle nuances.

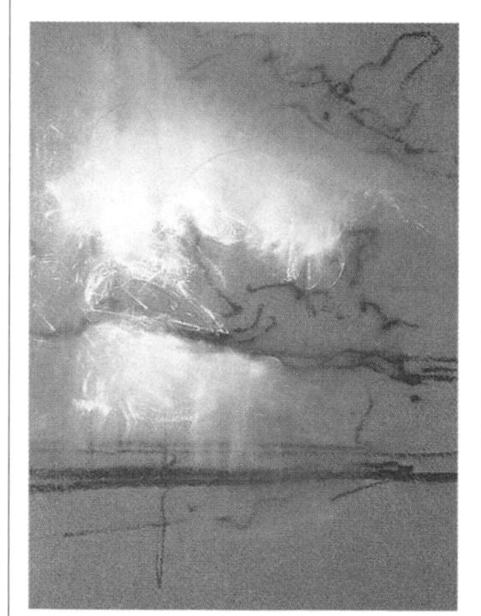

Edgar Degas, Marine-Soleil. *The dawn mist is masterfully expressed.*

Blending pastel colors.

A foggy landscape.

An enlarged detail of the background of landscape, showing the hills, its planes differentiated through color.

The atmosphere of a foggy landscape can be created using oil paints, particularly by the blending of different areas of color.

An enlarged detail of the foreground of the landscape. Notice the very general strokes for the sheep.

into-wet and wet-into-damp. By working the paintbrush the artist can guide color and moisture to where it best describes forms.

A Foggy Atmosphere Painted with Oils

This picture shows a foggy day on which the horizon has merged with the sky. The color of the sky is a grayish-blue mixed by using a lot of white and a little ultramarine blue, broken with a hint of ochre. For the most luminous parts of the cloudscape a little yellow is added to one of the base mixes.

To obtain the atmospheric effect of a foggy day, blends and tones are mixed using all the colors of the spectrum. These mixtures will also need to contain a lot of white. There are hues of green, blue, purple and a wide range of chromatically mixed grays.

The hills in the background have different chromatic casts within a range of similar tones. The contrasts between grounds are fairly subtle, although they are sufficient to distinguish between greater and lesser distances.

Remember that when working with oils, although the color white must be used to lighten color, it must be handled with care so that a clear gradation of values is maintained to create contrasts, however subtle.

MORE ON THIS SUBJECT

• Analyzing the Process **p. 50**

RAIN

When clouds turn to rain, water changes from gas to liquid. Creating the illusion of rain can be challenging. The effect of rain bears some parallel to mist, as edges are softened and colors modified. Brushstrokes and marks indicating texture and direction again take on importance.

Rain

How do you paint raindrops? We have seen that it is possible to draw a waterfall, since it has volume, motion, and direction. Raindrops are tiny, individual volumes of water that are barely seen. So how can they be represented?

A curtain of water, and the view of the landscape seen through it.

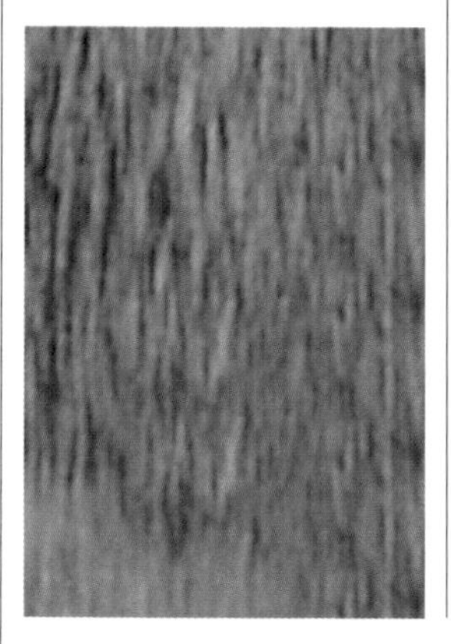

Clouds and pouring rain, using watercolor:

Initial, very dilute washes.

First a wash is applied with no suggestion of texture.

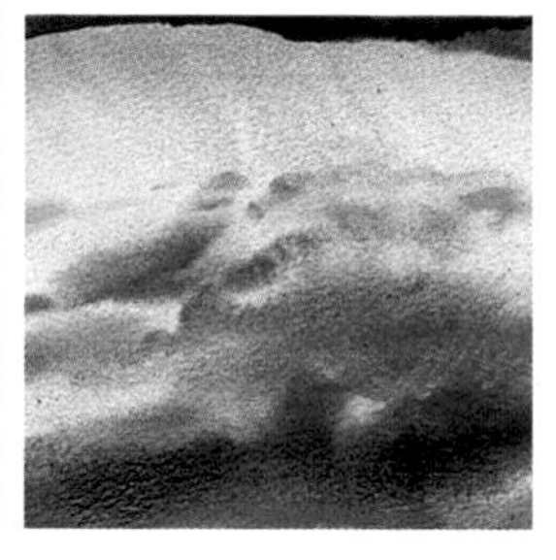

The artist searches for contrasts using Van Dyck brown and ultramarine blue, applied wet-into-wet.

The work of adding texture can be done with a brush.

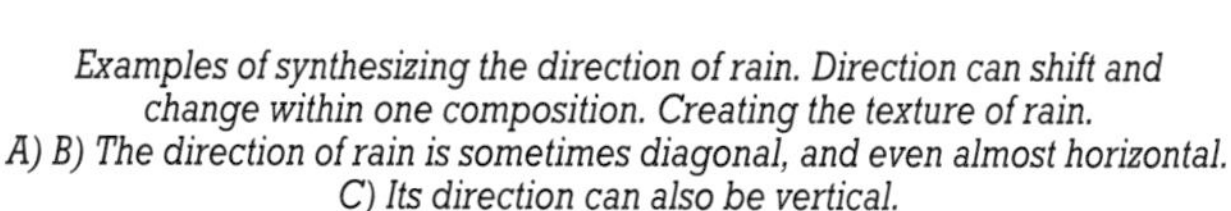

Examples of synthesizing the direction of rain. Direction can shift and change within one composition. Creating the texture of rain.
A) B) The direction of rain is sometimes diagonal, and even almost horizontal.
C) Its direction can also be vertical.

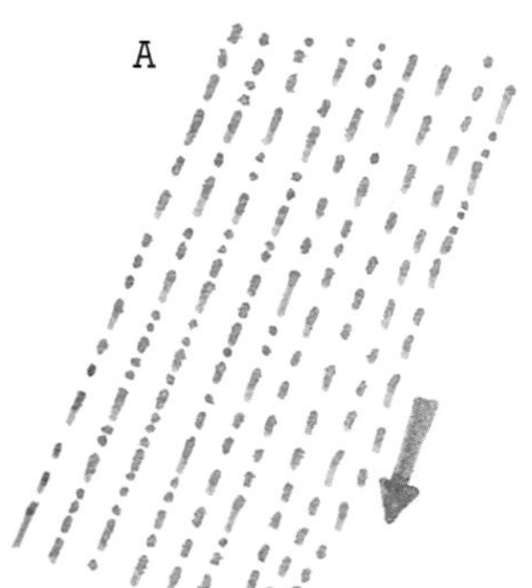

The visual perception of rain comes from the many points of ambient light held and reflected by its falling droplets. At the same time, the density of drops creates a mass, a thick curtain. The view of a landscape through this curtain is blurred. And rain has texture. Raindrops fall in a particular direction, carried by gusts that depend on the direction of the wind or straight down when there is no wind.

When painting a rain-swept scene one has to consider the density of the rain's curtain, the strength of ambient light that shines through it and that it reflects, and the direction of its fall.

Rain Using Watercolors

Watercolor has the capacity to represent a downpour with subtle and beautiful effects. Each new layer of paint is applied wet-into-wet. First a wash of cerulean blue is laid down to represent the lightest areas of the darkened sky. Subsequent washes will be darker, medium tones of Van Dyck brown and ultramarine blue. The painting is finished by adding areas of highly diluted cobalt blue.

This painting is worked entirely by brush. The paintbrush is used to color and draw the clouds. As the paint is applied wet-into-wet, the outlines bleed, so are soft and undefined. Other brushstrokes, also applied wet-into-wet, describe the direction of the rain. These strokes should be fine and intermittent, leaving some distance between each one. This will establish the direction of the rain and convey a curtain-like effect. See the illustrations on the preceding page, *Clouds and pouring rain, using watercolor.*

Rain using pastels.

MORE ON THIS SUBJECT

• Water: Texture and Rhythm **p. 44**
• Analyzing the Process **p. 50**

Rain Using Pastels

Strokes and points of color are added to a painted background to give it texture. These strokes show the direction of the rain. They are painted in colors that contrast with the background color. All the strokes and touches of color are irregular. Some of them represent small points of light.

The Texture of Rain: S'graffito

To depict the motion and texture of rain the artist can use also the technique of s'graffito. Pictured are various examples of coloring using different media: pastels, oil crayon, ink, watercolor, and oil. On each example, either dry or wet, s'graffito has been used on the surface using a sharp tool.

Examples of sigraffito:

A) Pastel. A palette knife is used to mark the color.
B) Oil crayon, incised with a pen nib or pointed object.
C) Ink, scraped with a razor blade or a mat knife.
D) Watercolor. This is scratched with a knife when the paint is dry.
E) Oil. The end of a paintbrush can be used to score grooves while the paint is wet.

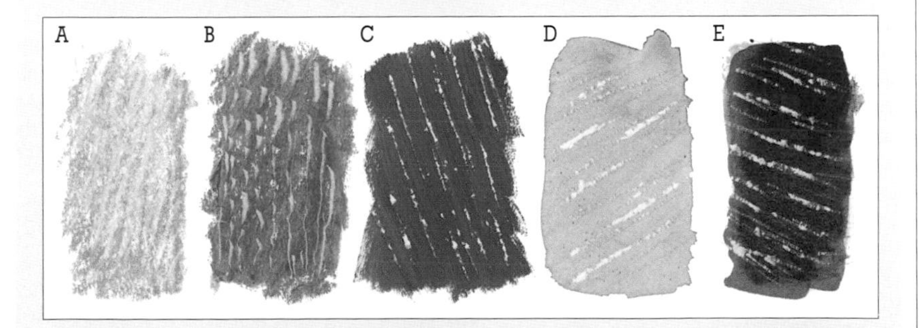

WET SURFACES, 1

In this chapter we examine some effects and techniques concerning the representation of light and color and reflections on surfaces made slick by moisture.

Reflections on Wet Sand

In the pastel work on the right, the interplay of the figure, its reflection by the sand, and the luminous, gray shadow it casts are important composition elements.

Besides mirroring the figure, the sand reflects ambient light. The sand's color becomes part of each mixture for the color of its reflections.

Reflections on Wet Streets

This urban landscape in the rain is done with oils. The first cool layer of color sets the tone for its chromatic scheme. The street and its shadows are painted next, using large, simple abstracted forms. The strong perspective of the street roots the abstraction in reality.

A clothed figure standing on wet sand at the water's edge.

The Color of Wet Earth

Dry earth is one color. The same earth when wet takes on a different chromatic cast. As a

A

Notice the colors in the reflections of the figure on the wet sand and of her projected shadow.

A) The reflections on the wet sand are a very well defined inverted image.

B) The color of the projected shadow conveys the wetness of the sand.

B

Gustave Caillebotte, Paris Street. A Rainy Day. *The artist's impression of reflections on wet streets.*

MORE ON THIS SUBJECT

- Reflections **p. 16**
- The Theory of Color **p. 18**
- Water: Line and Color **p. 42**
- Water: Texture and Rhythm **p. 44**

To depict reflections on wet streets it is important to lay down an initial coloring. This basic ground is then painted over and given mass and detail: An urban scene in the rain (left). Simple shapes and blocks of color to establish the main chromatic areas on a canvas previously prepared with a mixture of blues (right).

general rule, viewed in the same light, the color of damp earth is darker than when it is dry.

When it rains there is a change in the ambient light, which grows paler and more diffused, darkening the landscape. On rainy days the light has a silvery brightness that gives the landscape dramatic nuances. The colors in wet earth range from dark ochre, if chalky, to more reddish if it is composed of clay.

The color of wet earth.

The color of dry earth.

An Urban Landscape in the Rain

On a rainy day, the wet streets of a city have a general grayish tone. Palettes of broken grays and earth colors can be mixed using primary and opposite colors and white.

Reflections of pedestrians, cars, and buildings on the damp surface of a street can be strong composition elements in an urban landscape after the rain.

The ambient light is gloomy, everything tending towards bluish gray. Ambient light on shadows are reflected in the damp surfaces of the flagstones.

When working with oil, the initial coloring will set the chromatic tone and mood of the piece. Subsequent work of adding color, detail, and nuance describe the reflections on the damp surfaces of the streets. Reflections of ambient light create broad bright areas, the images of buildings are inverted.

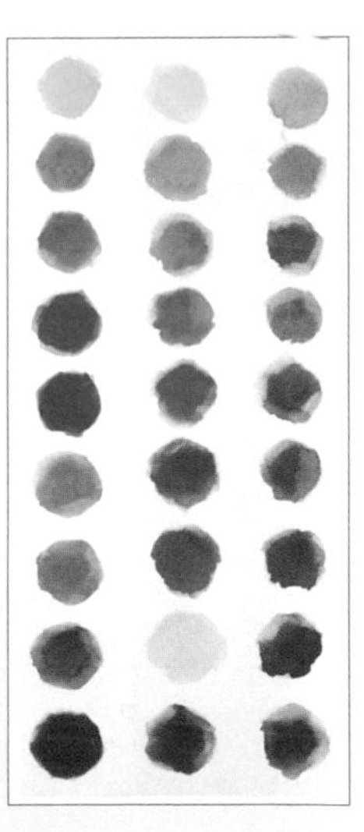

WET SURFACES, 2

When dealing with wet surfaces it is necessary to suit the medium to the task. In this chapter we will examine some representative examples of wet surfaces, particularly focusing on the figure and fabric, using watercolor and pastels.

Wet Skin Painted with Pastels

Wet skin, too, creates areas of reflection. The forms of the figure and the nature of skin causes shine to be highly accentuated. The skin's smooth, oily surface texture sometimes causes the formation of water droplets. By portraying these, as well as accentuating areas of reflection the painter will convey the impression of moisture.

Wet hair and wet fabric, painted with pastels.

Enlarged detail of the hair.

Wet skin painted with pastels. Enlargement of the details of the face of the male figure beneath the torrential rain.

Enlarged detail of the shirt adhering to the skin.

Wet Hair and Wet Fabric, Painted with Pastels

When hair or clothes are wet, they change color, form, and texture. Water reduces the volume of hair. Isolated locks form, falling downwards with the effect of gravity. Areas of intense shine appear, as well as a general darkening in color and tone.

When clothes are wet they also have a greater tendency to be pulled downwards under the effect of gravity. Creases are softened and fabric becomes transparent. In this pastel painting the shirt is transparent with moisture. The colors and tones of flesh can be seen through it. The forms of the figure can be glimpsed where the fabric adheres.

The blue of the pastel paper is used as a ground which has the character of blue reflections. The texture of the wet material, with its creases and the transparency which reveals the underlying skin is worked onto the blue surface. The color of pastel

Sorolla and Wet Skin

Wet skin produces more areas of intense shine than dry skin. Also, the general flesh tones are more intense. After a shower or a bath, the damp skin appears smooth and shining, with occasional isolated drops of water as well.

In this detail from the work *Children on the beach* by Joaquín Sorolla, the intense shine is bright white. The color of the flesh tones creates the illusion of glistening, tanned skin. The wave has just retreated and the reflections on the damp sand at the shoreline frame the figures. The composition is striped by light blue areas of water and reflection.

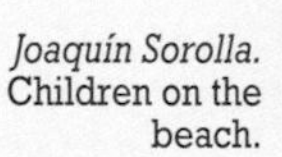

Joaquín Sorolla. Children on the beach.

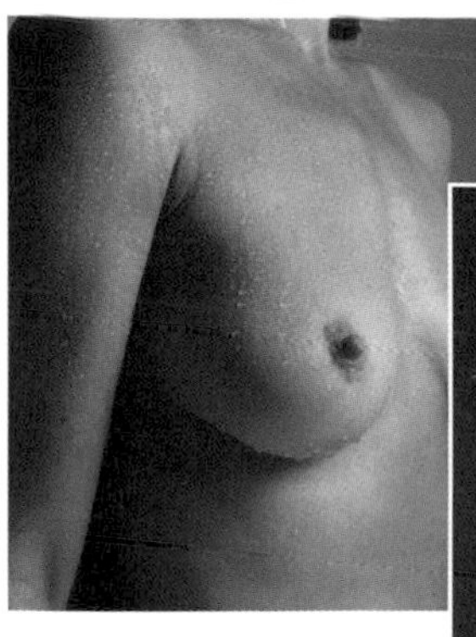

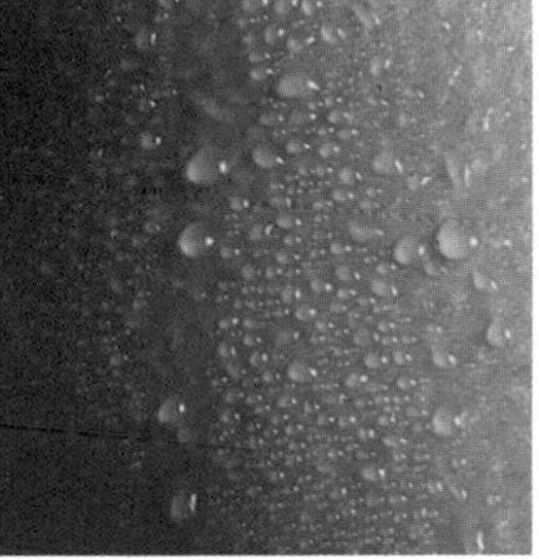

Figure. Nude with wet skin. Photograph and enlarged detail. The drops of water have points of light as well as shadows.

MORE ON THIS SUBJECT
- Reflections p. 16
- The Theory of Color p. 18
- Water: Line and Color p. 42
- Water: Texture and Rhythm p. 44

sets a chromatic mood and tends to unify a composition.

Painting Wet Skin with Watercolor

When using water colors to represent drops of dampness on human skin, masking fluid can be used to keep areas of paper clear. The fluid is used over a very diluted first wash that has been allowed to dry. These will be the points of greatest light. When the paint is dry, the rubber mask can be removed by rubbing it away with a finger. Then the color of the initial wash will appear. Through contrast with the darker washes, these become the points of light. The effect can be heightened by shadowing some of the water droplets using a paintbrush and a slightly darker wash. In the enlargement of the photograph, one can see many drops of water. But understatement in painting is often more effective than painting exactly what one can see. Only a few droplets need representation in order to convey a convincing illusion.

The initial wash and the application of masking fluid.

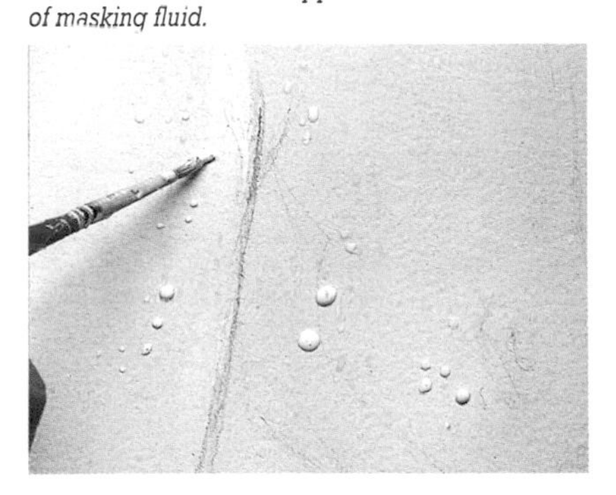

A later wash on the general tone of the base. The paint is allowed to dry and then the masking fluid is removed.

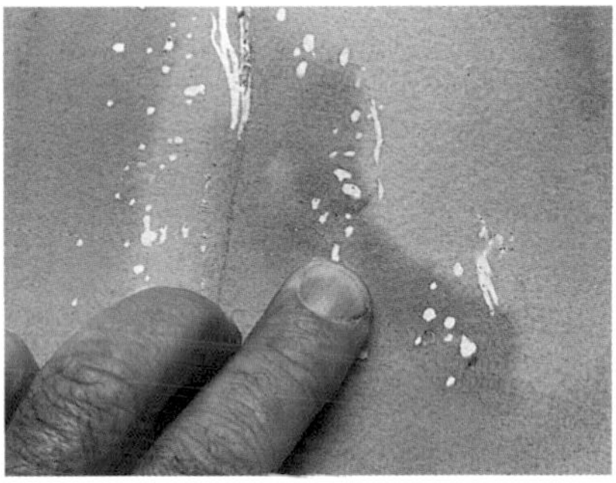

The volume is completed with more washes. The masked areas are given a lighter color than the general tone.

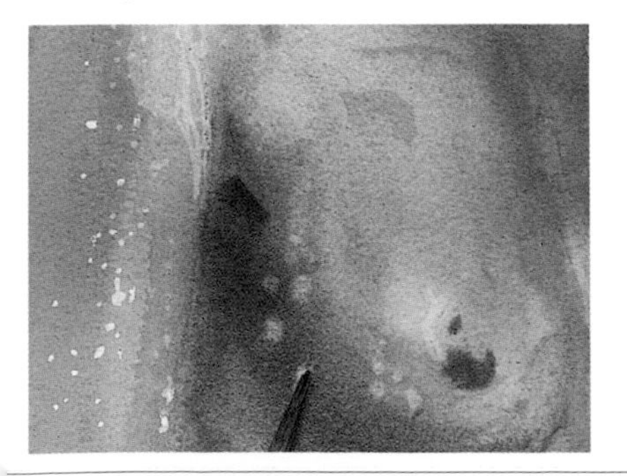

The effect of damp skin painted with watercolors can be further contrasted by shadows.

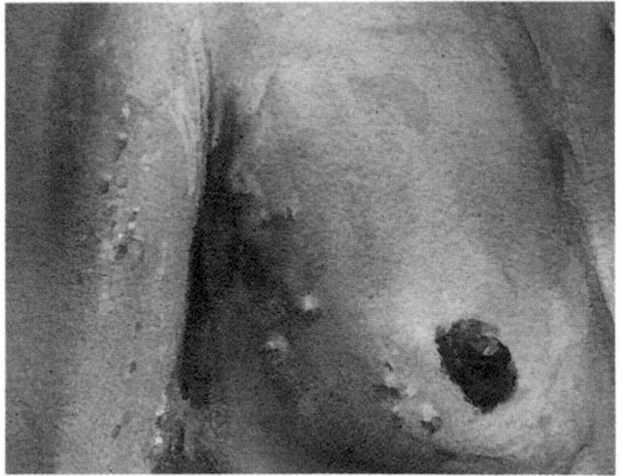

WATER AND LIGHT

Pictured is a series of masterpieces. The masters who painted them—Monet, Hoddler, Turner, Klimt, Burchfield, Homer, van Gogh, and Sorolla—have given us their various interpretations of water and light. Each is a powerful depiction, full of originality and vibrancy.

The Illumination of Water

The appearance of water in any of its states depends for the most part on ambient light. Nature affords infinite possibilities. Using personal vision, the painter captures water's different aspects, from the play of light and reflection on its surface, to chromatic harmonies and dynamic motion, to its dramatic contrasts.

Masters of Light

The rising sun, the light of dusk, nighttime moon, and starlight are the themes of great waterscapes. Let's explore this collection of the Masters.

Impression, Sunrise: Monet has captured a fleeting moment when dawn's blues are about to be transformed by the sun's orange light.

In his painting, *Starry Night on the Rhône,* van Gogh created a representation of a nocturnal panorama, with paths of golden starlight as reflections on the water.

Hoddler expresses the landscape of the Jungfrau with a cool surrealism. Light is the great protagonist in this landscape. His medium, watercolor, is transparent and luminous.

The light at dusk is one of the most magical moments of a day. Turner's magnificent seascape, *Sun Setting Between Dark Clouds,* is dramatic and simple. It too captures a fleeting moment, the sun scattering the dark night's murky blackness.

Claude Monet, Impression, Sunrise, *oil on canvas.*

Ferdinand Hoddler, The Eiger, The Mönch and the Jungfrau by the light of the moon.

Vincent van Gogh, Starry Night on the Rhône.

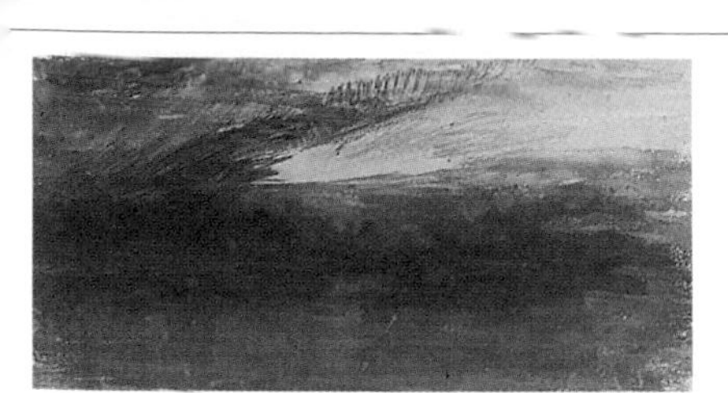

William Turner, Sun Setting Between Dark Clouds.

Charles Burchfield, Nocturnal Scene.

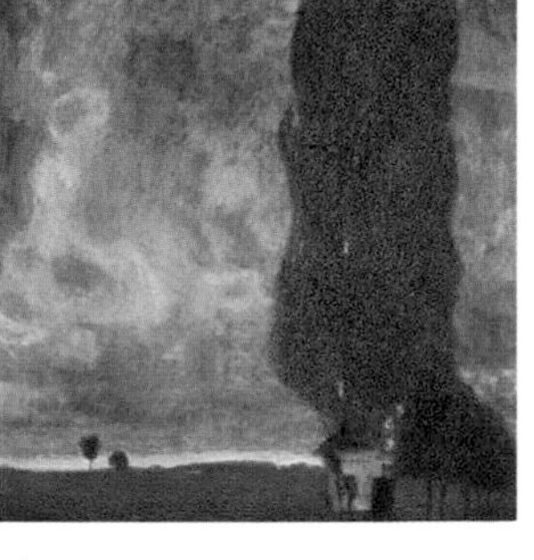

Gustav Klimt, The Great Poplar.

Sienna, Naples yellow, gray-blue, and black are the colors of its simple areas, providing a counterpoint for the intense, orange disk of sun.

In *The Great Poplar,* by Gustav Klimt, the painter depicts a deep, stormy sky that contrasts with the foreground, and is resolved in a flat, decorative way. The towering, lyrical shape of the poplar, and the flat ribbon of land, both painted in red tones, heighten the drama of the silvery blue-gray glow of the sky. The impact of light and form are powerful.

Nocturnal Scene, by Charles Burchfield, transmits a great feeling of dramatic strength, using energetic, descriptive brushstrokes, and a surrealistic depiction of stars.

Gloucester Sunset, by Winslow Homer is painted in blues and oranges. It is full of powerful contrasts, awash in its warm, orange light. Floating on the bright surface of water, the boats are only silhouettes.

Winslow Homer, Gloucester Sunset.

MORE ON THIS SUBJECT

- Reflections **p. 16**
- The Theory of Color **p. 18**
- Water: Line and Color **p. 42**
- Water: Texture and Rhythm **p. 44**

Contrasts of Light and Shade

Joaquín Sorolla is known as a painter of light. With this work, *Storm,* he demonstrates his mastery of the palette and his preference for themes of strong light and contrast. Three basic color areas make up its beautiful composition.

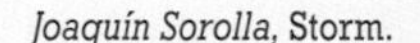

Joaquín Sorolla, Storm.

WATER AND WATER CONTAINERS

Stationary water must be contained. Because of gravity, the sea is held by the earth's depths, its edges the shores and cliffs. In smaller quantities water is contained by and assumes the form of its container.

Outline

Water assumes the shape and outline of the object which holds it. Illustrated here are various receptacles containing water. The liquid has a flat surface, while the shapes of the containers define its other borders.

The Color of Water in a Container

The color of water as it is seen in a container depends first of all on the color of the container. The color of the glass or plastic will give the water its own coloration.

Distortion of the Outline

Because of the transparent or semitransparent surface of the container, and perhaps its variation of thickness, the outline of the water usually appears distorted.

In the two still life paintings on these pages, a colored liquid has been represented in clear containers.

One depicts a container of orange juice. The juice behaves as water does, only varying in its orange color.

The outline of the container alters the forms. Notice the distortions and variation of color produced by the ridges of the glass. The artist has further exaggerated these effects which heightens the dramatic impact of the object.

Containers of different shapes containing water. In one of them the water is clean and clear. In the other the water is dirty. Notice the impact this has on the projection of shadows and transparencies. The color of the container affects the color of the water contained in it.

Using Pastels

Let's study the representation of milk contained in this glass and pitcher, using pastels.

The choice of color of Mi-Teintes paper is the first important decision in this exercise. We have used a neutral color, chosen from a wide range of colors available for paper produced by Canson.

Using pastels: the subject and its representation.

The still life is resolved using two colors of pastels, white and one sepia, on a Mi-Teintes colored paper.

Container containing orange juice.

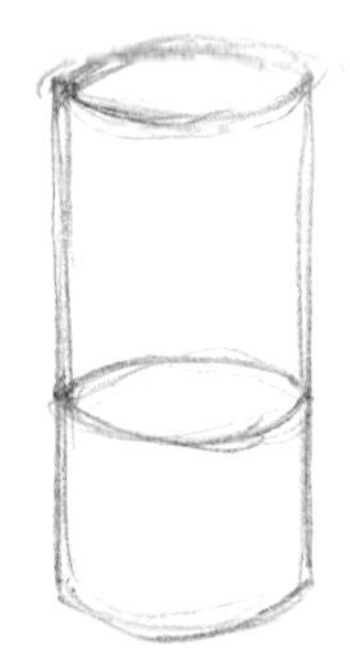

A sketch for the container of juice.

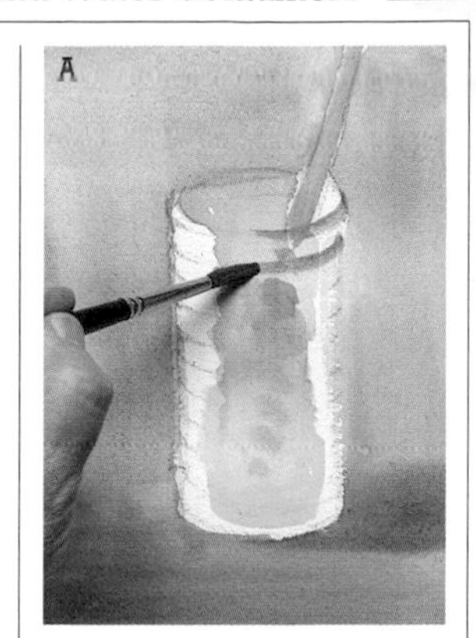

Using watercolors: A) First washes. B) Adding detail with successive washes. C) Final details and contrasts.

It provides a unified color ground, allowing the artist to achieve a general, overall color intonation very quickly.

In this painting, the artist, Ballestar, has used only a few colors. A gradation of grays models the forms of pitcher and glass. The milk contained appears opaque, while the glass objects have the convincing illusion of clarity. With their folded napkin, these objects make up a lovely composition in black, white, and gray.

Using Watercolors

The orange juice tumbler is painted with watercolor. It owes its sharp image to the careful sequence used to develop it.

With the goal of achieving vibrant colors, the first coat of watercolor is applied. Areas of white are carefully left untouched to depict the reflections and bright parts of the container.

Contrasting colors for the outline of the container are applied with a brush.

The distorted shape of the stirrer is also brushed in.

Finally, the work takes on the finished expression of volume.

MORE ON THIS SUBJECT

• Water: Line and Color **p. 42**

Image Distortion Through Transparency

If one looks at an object through a glass container filled with water, the image one sees is not clear. The distortions or blurred appearance of the object depend on the thinness and thickness of the glass and on the clarity of the water. But glass invites chromatic and exaggerated depictions, and the painter can embellish the work with rich variations.

Image distorted through transparency. Through the essentially transparent vase and water, the image of the stalks looks distorted.

Original title of the book in Spanish: *Agua*

Published by Parramón Ediciones, S.A., Barcelona, Spain.
Author: Parramón's Editorial Team
Illustrators: Parramón's Editorial Team

All inquiries should be addressed to:
Barron's Educational Series, Inc.
250 Wireless Boulevard
Hauppauge, New York 11788
http://www.barronseduc.com

International Standard Book No. 0-7641-5162-2

Library of Congress Catalog Card No. 99-62173

Printed in Spain

9 8 7 6 5 4 3 2 1

Note: The titles that appear at the top of the odd-numbered pages correspond to:

The previous chapter
The current chapter
The following chapter